Innovative and Efficient Laboratory Management

Innovative and Efficient Laboratory Management

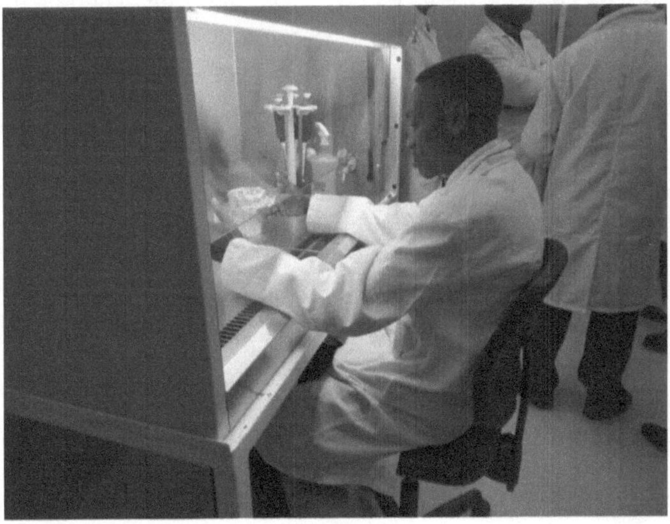

Godswill Ntsomboh Ntsefong

ISBN: Softcover 978-1-5035-6437-4
eBook 978-1-5035-6436-7

Print information available on the last page.

Rev. date: 04/23/2015

To order additional copies of this book, contact:
Xlibris
1-888-795-4274
www.Xlibris.com
Orders@Xlibris.com
700677

Dedication

I dedicate this work to my beloved father Nde Fomenop Thomas Ntsefong, whom I know would have been elated if still alive today, to see and appreciate the fruits of the precious moral and ethical education he gave me as heritage.

Preface

Within the last two decades, the phenomenon of management has acquired a remarkable visibility. The challenges of organizational management are directly related to economic growth, social development and the policy choices of societies. By developing this theme on the laboratory, a structure for research, diagnosis, prevention or production, the author recognizes it as a context where the future/wellbeing of people is designed and decided. Meanwhile, there are real gaps between technical knowhow and management, as well as between general management and laboratory management, the general objectives of laboratories must be attained for their activities to keep running smoothly and for their efficient contribution to the quality of life. The focus on the laboratory as an entity comprising all specialties is pertinent of this book.

Technologist, laboratory technician, operator, and laboratorian, are some designations of those who work day and night in various scientific domains. Usually considered as executors, their efficacy is limited when it comes to planning, organizing, directing and evaluating. The common perception is that such tasks are reserved for others and that being a technical specialist in the laboratory signifies no need of seeing beyond…the objectives of a microscope. Such duties are therefore trusted to others like trained managers or administrators whose basic training has nothing to do with the specified scientific domain. Such leaders find it difficult to make technical specialists understand the importance of putting in place and/or respect of policies, procedures and their application. This results to less implication of the latter, consequently leading to frustration, misunderstanding, wastage, supply shortages, equipment failures, erroneous or less authentic data or results, and bad reputation of laboratories.

The author has met these difficulties through his scientific journey. His interest in research is focused on Plant Biotechnology with a special objective of improving dietary oil quality. The high quality training he received on good laboratory practices and management principles has completely changed his vision on how to work as a laboratory manager. In this book, he puts together information that has been most useful for him to improve on his efficacy. His view is not limited on a precise type of laboratory but covers all scientific domains. *Innovative and efficient laboratory management* treats the laboratory from a global perspective: biosciences, industrial, and clinical research. The presentation is practical and can constitute a solution to all types of managers and laboratory personnel. From administrative organization to Good Laboratory Practices via Biosafety and Biosecurity and the notion of Quality Control, Quality Assurance and ethics in research, all is well set to give the reader a clear understanding of the importance of laboratory recognition through the process of accreditation.

The book is actually an innovative solution to the problem of laboratory efficiency and can constitute the basic competence of the laboratory. It can help to manage a laboratory project or develop a private or public career, while re-orientating or equiping in the jungle of laboratory careers. This document can valuably serve as course material for lectures in scientific programs of training institutions and universities, in order that the "management culture" is inculcated right at the foundation of scientific training.

Are you a manager, biologist, technician, trainer or lecturer? Discover the content of "*Innovative and efficient laboratory management*" and progressively 'unveil the blind' that has so far prevented you from being efficient in the administrative organization of your laboratory, or in Good Laboratory Practices, Biosafety and Biosecurity, management of equipment, reagents and supplies, or in putting in place of a Quality System and the practice of Quality Control. The content can equally lead to an engagement in a sure process towards accreditation as the ultimate formal recognition of the quality of services of your laboratory.

Georgette NDONGO NKENG épse EKANGA
Biologist/Specialist in Management of Health Services
Chief of Medico-Sanitary Techniques Services
University Teaching Hospital (CHU) Yaounde, Cameroon

Abbreviations

ABSA: American Biological Safety Association

AEL: Airborne Exposure Limits

AMR: Analytical Measurement Range

APLAC: Asia Pacific Laboratory Accreditation Cooperation

BATA: Biological Agents and Toxins Act (Singapore)

BBFC: Bioscience-Bioethics Friendship Co-operative

BecA-ILRI Hub, Nairobi – Kenya

BLQS: The Bureau of Laboratory Quality and Standards (Thailand)

BMBL: National Institutes of Health Biosafety in Microbiological and Biomedical Laboratories of the United States of America.

BSE: Bovine Spongiform Encephalopathy (mad cow disease)

BSL: Biosafety Level

BW: Biowarfare

CDC: Centers for Disease Control and Prevention (USA)

CEREPAH: CEntre specialisé de REcherche sur le PAlmier à Huile de La Dibamba

CFRs: Codes of Federal Regulations

CJD: Creutzfeldt Jakob disease

CRR: Clinically Reportable Range

CWM: Clinical Waste Management

DRID: "Deliberately Reemerging Infectious Disease" (bioterrorism)

EQAS: External Quality Assessment Schemes

GCLP: Good Clinical Laboratory Practice

GLP: Good Laboratory Practice

GMM: Genetically Modified Microorganism

GMP: Good Manufacturing Practices

HIV: Human Immunodeficiency Virus

IEC: International Electrotechnical Commission

ILAC: International Laboratory Accreditation Cooperation

IP: Intellectual Property

IQC: Internal Quality Control

IRAD: Institute of Agricultural Research for Development, Cameroon

ISO: International Standardization Organization

IT: Information Technology

KAN: National Accreditation Committee (Indonesia)

LAIs: Laboratory acquired infections (e.g. SARS, Ebola, vaccinia)

LIMS: Laboratory information management system

LIS: Laboratory Information System

LSDG: Laboratory Standard and Design Guide (Stanford)

MOHSW: Lesotho Ministry of Health and Social Welfare

MRA: Mutual Recognition of Accredited results

MSDS: Material Safety Data Sheets

MSW: Maintenance, Support, and Warranty

MTA: Material Transfer Agreements

NABL: National Accreditation Board for Testing and Calibration Laboratories (India)

NJDHSS: New Jersey Department of Health and Senior Services

OSHA: Occupational Safety and Health Administration of the U.S. Department of Labor

P&P: Laboratory process and procedure

PEL: Permissible Exposure Limit

PEOSH: Public Employees Occupational Safety and Health (New Jersey)

PPE: Personal Protective Equipment

PSDS: Product Safety Data Sheet

QA: Quality Assurance

QAU: Quality Assurance Unit

QC: Quality Control

QI: Quality Improvement

QM Quality Manual

QMS Quality Management Systems

R&D: Research and Development

RIMS: Research and Innovation Management Services bvba

SA: Select Agents

SaaS: Software as a Service

SARS: Severe Acute Respiratory Syndrome virus

SARS-CoV: Severe Acute Respiratory Syndrome Corona Virus

SC: Supply Chain

SCM: Supply chain management

SDS: Safety Data Sheet

SEA: South East Asia

SLMTA: Strengthening Laboratory Management Towards Accreditation programme

SOPs: Standard Operation Procedures

SRs: Russia's Sanitary Rules

TQM: Total Quality Management

TSE: Transmissible Spongiform Encephalopathy

UNEP: United Nations Environment Programme

VFM: Value for money

WHO LBM: World Health Organization Laboratory Biosafety Manual

WHO: World Health Organization

WHO–AFRO–SLIPTA: World Health Organization Regional Headquarters for Africa's Stepwise Laboratory Quality Improvement Toward Accreditation

WLC: Whole Life Costing

Table of Contents

Dedication v
Preface vii
Abbreviations xi
Acknowledgement xix
Introduction xxi

Chapter 1. Laboratory or Test Facility management basics 1

Safety Training 2
Laboratory administration 2
The Laboratory Organigramme 3
The Laboratory Manager 4
The Quality Manager 7
The IT Manager 8
Documentation Officer 8
Financial Officer 8
Chief Laboratory Technician 9

Chapter 2. Laboratory design considerations 10

Objectives and preconditions 10
Safety issues in Laboratory building design 11
Safe access 13
Sanitation considerations 14
Electrical and plumbing considerations 15

Chapter 3. Good Laboratory Practices 16

Use of SOPs and Quality Manual 17
Use of Material Safety Data Sheets 19

Labeling as a GLP 20
Work practice and engineering controls 20
Use of Personal Protective Equipment 21
Establishing a PPE program 23
Categories of PPE 24

Chapter 4. Biosafety and Biosecurity 27
The rationale for Biosafety and Biosecurity 27
Risk assessment for Biosafety and Biosecurity 29
Laboratory Biosecurity 31
Brief on the American Biosecurity experience 33
Laboratory Biosafety 33
Biosafety potentials for consideration as a separate academic discipline 33
Laboratory classification into Biosafety Levels 35
BSL-I Laboratory 36
BSL-II Laboratory 36
BSL-III Laboratory 36
BSL-IV Laboratory 37
Laboratory classification based on context and threat 38
The SARS-CoV experience 38
Management of Laboratory waste as a GLP 40
Historical background of waste management 40
Laboratory waste management 41
Environmental impacts of Laboratory wastes and operations 42

Chapter 5. Procurement of Laboratory Supplies 44
Laboratory goods and services 44
Laboratory supply chain management 45

Chapter 6. Laboratory equipment operation 47
Categories of Laboratory equipment 47
Laboratory equipment maintenance 49

Chapter 7. Laboratory information management system 52
Core functionalities associated with LIMS 53
LIMS architectures 55
Thick-client LIMS 55
Thin-client LIMS 56
Hybrid LIMS architecture 57
Web-enabled LIMS architecture 57
Web-based LIMS architecture 57
Some limitations of LIMS 58
A case study of LIMS 59

Chapter 8. Quality Control and Quality Assurance 61
Quality Control 61
Quality system 62
Quality assurance 63
Total Quality Management 64

Chapter 9. Laboratory Accreditation 67
Accreditation process essentials and requirements 68
Implementation of accreditation at national level 69
National accreditation standards, a step towards conformity with international standards 70
International standards for Laboratory accreditation 72
Benefits of Laboratory accreditation 76

Chapter 10. Ethical issues in research management 77
Some strategies to implement ethics 80
Material transfer and ethics 81

Conclusion 83
References 85

Acknowledgement

This book will not have come to existence if several people and institutions did not help me in one way or the other. I am grateful to the BecA-ILRI Hub, Nairobi – Kenya (and their partners) from where I received initial training on Laboratory management and equipment operations, and without which the idea of this write-up might never have crossed my mind.

My hearty gratitude also goes to the General Manager of the Institute of Agricultural Research for Development, IRAD Cameroon, Dr. Noé Woin, and the IRAD hierarchy for appointing me as Laboratory Manager, a responsibility that has edified me on the actual implementation of certain concepts presented in this book. Particular thanks go to Dr. Achukwi Mbunkah Daniel, Scientific Director of IRAD Cameroon, who believed in my potentials and capability of coming up with a Laboratory guide for my research colleagues. The said guide has actually boosted me and enhanced the effective realization of this book.

I also thank Dr. Koona Paul and Dr. Ngando Ebongue G. F., respectively former and actual Managers of the Specialized Oil Palm Research Centre (IRAD-CEREPAH of La Dibamba) for their scientific mentorship and guidance on my job as a researcher and Laboratory Manager. I also thank my beloved brother and colleague, Maho Yalen Edson for taking his precious time to read through and comment on the original manuscript and with whom I have often discussed about some of the ideas leading to the realization of this work.

My gratitude to all students currently under my mentorship and on internship in the Lipids Analysis Laboratory of IRAD Dibamba for their inspiration towards effective implementation of GLP within the Laboratory. I think of Sastile Ngangue, Kenmogne Simo Thierry, Di-Maissou Jean Albert, Tabi Mbi Kingsley, Likeng-li-Ngue Benoit, Epoh Nguea Toussaint, Kwembi Aimée, and Pam Opportune. Dear friends, note that what we do and share on a daily basis have really contributed to the realization of this book which I hope will serve you as an example to emulate in your research careers. I am very grateful to Kooh Jeanne of IRAD Dibamba who supported me immensely with her laptop computer during the last phase of this book when my laptop suddenly stopped working. All my dreams and deadlines for the effective submission of the manuscript to the publisher almost went into oblivion just before she stepped in. May God bless you my sister. My sincere gratitude goes to Mrs. Georgette Adèle NDONGO NKENG epse EKANGA, Clinical Biologist and Manager of Health Programs at the University Teaching Hospital (CHU) Yaounde, Cameroon, for accepting to review the original manuscript and produce a preface to this book. May God increase her wisdom.

Finally, I am indebted to my beloved wife [Melanie Celestine] for her constant encouragement and inspiration. Thanks also go to my children [Asriel, Kadmiel, Lael, Magdiel, Mabel, Adel and Daniel] for giving me enough space at home and sufficient time off from home which I much needed to concentrate and come up with this final product. May this work serve them as proof or as the result of hard work and perseverance! To all those who contributed in one way or the other to the success of this project and whose names are not cited here, I invoke God's abundant blessings!

Godswill Ntsomboh Ntsefong
La Dibamba
April 10, 2015

Introduction

Good research and Laboratory management practices fall in line with industrial, clinical and agricultural challenges which can be addressed efficiently with scientific tools notably for such issues as service and product quality, pests, diseases and climatic constraints that result in poor environmental and human health, low crop yields and poor animal productivity. Such major challenges could be solved efficiently if opportunities linking industrial research and development and modern biosciences to environmental, health and agricultural improvement were exploited appropriately. Food security for instance could be assured through research in crop, microbe and livestock areas where exploration of new developments in biosciences and proper research and Laboratory management are imperative prerequisites. Since most research activities involve Laboratory work, there is need for efficient management of Laboratory or test facilities to ensure quality controlled research and cost effective use of resources. For some years now, this has been the effort of the BecA-ILRI Hub (Kenya) from where the author received training on principles in Laboratory management and equipment operations. This book was initially inspired by the need to restitute lectures received from the said training to fellow research colleagues back home. The idea was further developed through literature review and hands-on experience of the author as Manager of the Lipids Analysis Laboratory of the Institute of Agricultural Research for Development (IRAD) in Cameroon.

It is obvious that good Laboratory and research management skills are necessary for scientists and scholars involved directly or indirectly in industrial, clinical or bioscience research and/or charged with management of Laboratory facilities. The essence of this write-up is therefore to enhance good Laboratory management practices that ensure stressless compliance with legal and regulatory frameworks for health and safety. *In fact, moral distress occurs when one knows the ethically appropriate action to take but is constrained from taking that action.* Our aim is to promote scientific excellence by highlighting the conditions and skills necessary for efficient and innovative management of Laboratory facilities while enhancing consciousness and efficacy in cost effective research management.

The issues addressed in the first chapter of this book include the definition or presentation of the Laboratory or Test Facility and the administrative setup which involves the persons, premises and operational units that are necessary for conducting research safely. The second chapter dwells on Laboratory design considerations which notwithstanding have a significant impact on the quality of results generated. In fact, accurate results are obtained confidently from a safe environment for Laboratory personnel. However, no matter how well designed a Laboratory is, education of the facility users is essential since improper usage of its facilities will always defeat the engineered safety features.

In the third chapter, we present a quality system concerned with the organization, process and conditions under which tests are conducted. This quality system is termed Good Laboratory Practice (GLP) which is a set of rules, operating procedures and the proper practices that ensure the reliability of the data generated by laboratories. The principles or parts of Good Laboratory Practices are: organization and personnel, installations and facilities, documentation, equipment and instruments, materials and reagents, reference and assay samples, assay methods' validation, etc. This chapter reiterates the importance of efficient implementation of GLP involving labelling of equipment, materials and reagents, the establishment and effective use of Standard Operation Procedures (SOPs), effective use of Laboratory Quality Manual (QM) and Material Safety Data Sheets (MSDS), work practice and engineering controls, and the use of personal protective equipment (PPE) in the Laboratory for safety.

The fourth chapter elucidates the importance of Biosafety and Biosecurity, two components of Good Laboratory Practice needed to counteract the existing threats especially with regards to the use of bioagents in terrorism. Biosecurity encompasses physical and administrative measures designed to protect biological material like pathogens, toxins, and sensitive information that could cause harm to health or economic loss resulting from malicious release, intentional loss, theft, sabotage, or misappropriation, and subsequent misuse. Meanwhile, Biosafety is the reduction and prevention of the risk of exposure of workers and the environment to dangerous pathogens or hazards. Laboratory waste management and the classification of Laboratories into Biosafety Levels are also treated in this section.

Procurement of Laboratory Supplies is treated in chapter five. Here, the author reiterates the importance of a procurement strategy for each Laboratory or test facility, whose aim should be to set out a planned approach for cost-effective purchasing of required goods and services, taking into account several factors such as the timeline for procurement, the funding and budget and the projected risks and opportunities. The chapter upholds the need for a defined and documented policy and procedures for selection and use of purchased external goods and services in addition to an inventory control system for Laboratory supplies. The goal here is to achieve Value for money (VFM) in the Laboratory supply chain (SC) by obtaining the right (quality) goods and services, with timely delivery and at the right cost.

Chapter six deals with Laboratory equipment operations. Laboratory operators need to have an overview of different categories and types of Laboratory equipment at their disposal with good knowledge of their safe handling, operation and maintenance following well set schedules. Meanwhile, Chapter seven dwells on Laboratory information management system (LIMS) which is an important and integral part of Laboratory operations relevant for efficient Laboratory management. The next chapter (eight) treats Quality Control (QC), Quality Assurance (QA) and Total Quality Management (TQM) as the three major elements of quality.

This chapter sets the stage for Laboratory accreditation in the next chapter (nine) which culminates all efforts of efficient Laboratory management expressed through the concepts developed in the various sections of this book. Laboratory accreditation which demonstrates legitimacy and credibility of results is the process by which an independent and authorized agency certifies the quality system and competence of a Laboratory on the basis of certain pre-defined standards. It is the formal recognition, authorization and registration of a Laboratory that has demonstrated its capability, competence and credibility to carry out the tasks it is claiming to be able to do. In this chapter, the reader will discover the whole process of Laboratory accreditation with the various agencies involved as well as the benefits of Laboratory accreditation.

The last chapter (ten) deals with ethical issues in research management. It is obvious that the consideration of ethics in research should enhance mature decision-making in harmony with changing technology. The chapter points out the fact that efficient research and Laboratory management must be based on ethical principles that guarantee all stakeholders' access to the benefits of new technologies with increased understanding of biological systems and responsible use of technology. This book closes up with a brief conclusion outlining some basic guidelines on how to implement knowledge gained from the book to efficiently run a modern Laboratory or research facility.

Chapter 1

Laboratory or Test Facility management basics

The relevance of efficient research and Laboratory management is based on the growing scope of international cooperation and exchange of work between scientists, coupled with the increasing threats from the use of dangerous pathogens by terrorists and the need for appropriate biological safety conditions for work, storage and transportation of these agents (Stavskiy *et al.,* 2003). Globally, Laboratory testing or research is undertaken in universities, clinical or diagnostic facilities, industry, government facilities, and other entities. Laboratory or Test facility activities can be identified as basic research, applied research, and clinical or diagnostic work (Gaudioso *et al.,* 2006) and those directly involved in such activities include Biosafety officers and responsible Officials, Principal Investigators, scientists and technicians, Directors or Managers, and Laboratory support staff and technicians (Gaudioso *et al.,* 2006).

In fact, a Test Facility includes the persons, premises and operational units that are necessary for conducting the non-clinical health and environmental safety studies (Isin Akyar, 2011). The term "test facility" may include several "test sites" at one or more geographical locations, where phases or components of a single overall study are conducted. It does not only include buildings, rooms and other premises, but also the people who are working there and are liable for performing these studies (Seiler, 2005; Isin Akyar, 2011). For multi-site studies, the test facility considers the site at which the Study Director is located and all individual test sites, which individually or collectively can be considered to be test facilities. The different test sites in the test facility may include Research Laboratories where test or reference item characterization considering determination of identity, purity and/or strength, stability, and other related activities are conducted, and one or more agricultural or other in- or outdoor sites where the test or control item is applied to the test system (Isin Akyar, 2011). In some cases, Test facility can be a processing unit where collected commodities are treated to prepare other items or where collected specimens are analyzed for chemical or biological residues, or are otherwise evaluated (OECD, 1998; Isin Akyar, 2011).

Safety Training

The Laboratory Manager and all Laboratory staff must receive safety training. This should be well documented and completed before any employee begins working in the laboratory and on a regular basis thereafter. Such training should include issues such as Blood borne pathogens, PPE, Chemical Hygiene/Hazard Communications, use of safety equipment in the laboratory, use of cryogenic chemicals (e.g. dry ice and liquid nitrogen), transportation of potentially infectious material, waste management/biohazard containment, and general safety and/or local laws related to safety (Ezzelle *et al.,* 2008).

Laboratory administration

A safe and healthful test facility or Laboratory environment is the product of individuals who are trained and technically proficient in safe practices (Tun *et al.,* 2007). This section presents a proposal of how the Laboratory should be managed. The proposed organigramme (fig. 1) could be adapted or readjusted by Laboratories depending on their size or kind or activity. Our recommendation is that whatever the case, there should be a consistent team in place which reports to the institution's administration through the Laboratory Manager on a regular basis for the efficient running of the Laboratory. This Laboratory Management team could be constituted of the Laboratory

Manager, a Quality Manager, a Financial Officer, an Information Technology (IT) Officer, a Documentation Officer and a Laboratory technical staff Leader of Chief Laboratory Technician.

The Laboratory Organigramme

Efficient Laboratory management requires that the Management appoints a Laboratory or Technical Manager who has overall responsibility for the technical operations and provision of resources needed to ensure the required quality of Laboratory procedures. For a big Laboratory, the Laboratory Manager could be assisted in his duties by a Quality Manager, a Financial Officer, an IT Officer and a Document Control Officer. The Quality Manager should have the responsibility and authority to oversee compliance with the requirements of the Laboratory's quality management system. The Quality Manager could report directly to both the Laboratory Manager and the top management of the institution in which the Laboratory or test facility is based. On his part, the Document Control Officer could be identified and assigned control over all the documents received or developed within the Laboratory setting (Panadda, 2007).

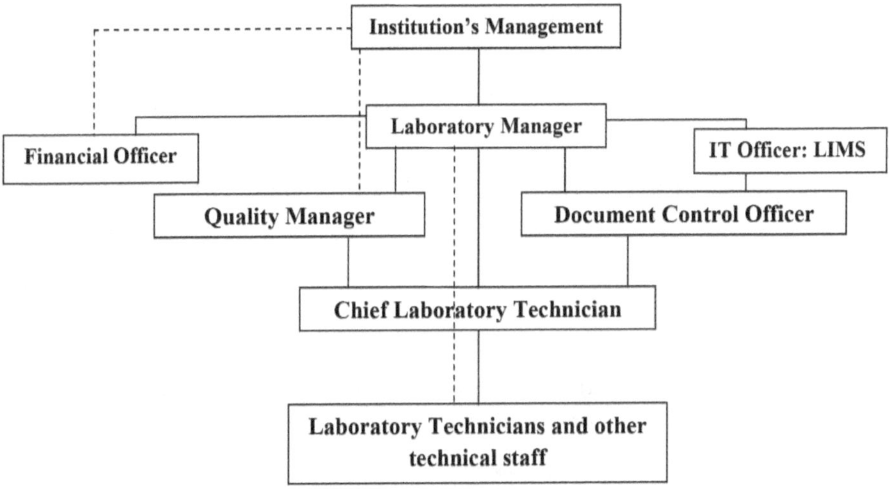

Figure 1. A simple Laboratory Organigramme
[Author's construction]

The Laboratory Manager

Figure 1 is a simplified organigramme of a Laboratory administrative setup. Hard lines indicate in this figure indicates that the officer below directly reports to the immediate hierarchy who oversees all activities of the officer under his/her direct administration. Meanwhile, the broken lines show that an officer though reporting to the direct hierarchy may be expected to be accountable to an even higher authority in the ranking. These broken lines equally indicate that the Laboratory Manager has an overview of all units of his/her administrative setup.

Regarding the role of the Laboratory Manager (figure 1), he is responsible for managing and coordinating the non technical activities while assuring a safe and well-run Laboratory. Since the Laboratory Manager or Head of the Laboratory is responsible for work related to instructing, managing, advising and training in the Laboratory, it is important for him/her to have a relevant academic qualification and if possible a professional technologist's license with adequate work experience (Panadda, 2007). The Laboratory Manager's potentials needed in this context should be inspired by his/her creative spirit in guiding or leading his team through good planning towards the achievement of a goal; that of efficient and innovative management of the Laboratory. He must therefore exercise leadership that takes into consideration the ethical, aesthetic, moral, intellectual, and environmental aspects of the Laboratory and his followers or the entire staff.

It could be interesting for the Manager to establish a Laboratory setting that enhances the development and improvement of competencies for effective organization, motivation and proficient output. The Manager needs to have a good mastery of the organizational setup, its goals and the resources available while constantly developing his/her potentials of managing with objectives. He needs to have a nurtured mind poised for growth and open to criticism. By this, he will be able to develop and balance critical competences at the individual, interpersonal, team, and organizational levels through formal training and practice. Even after many years of practical experience, there is always a need to update laboratory knowledge through written information, courses and exchange visits to gain practical knowhow (Geir *et al.,* 2011).

The Laboratory Manager who may or may not be the institution's Principal Investigator (PI) needs to stay current on up-to-date information and critical feedback on research practices by using multiple resources including journals, conferences, e-mail, and the Internet (Gaudioso and Zemlo, 2007). In order to be at the forefront with a broad view of research trends, and a focus for the lab, the Laboratory manager must follow the evolution of the literature and identify research trends to gauge the research topic areas that (1) have received the most scholarly attention over the years, (2) have fallen out of favor and have become historic research topics, and (3) have fallen out of favor, but have re-emerged, and have become resurgent research topics (Jeffrey and Stefanie, 2009). This could help to shape the line of research of the Laboratory and avoid producing out-of-fashion research findings. Moreover, the Laboratory Manager must ensure that all experiments performed in the Laboratory are based (when necessary) on well developed and validated methods. Method development and validation within the Laboratory setting is an essential aspect of credibility and data validity which helps to demonstrate performance specifications comparable to those established by the manufacturer to ensure that assays are performing optimally within the proposed testing environment. In this context, documentation of experimental results and approval should be readily accessible (Ezzelle *et al.,* 2008). Verification and documentation of normal responses for each test system including normal, Analytical Measurement Range (AMR) or the Clinically Reportable Range (CRR) must be established to determine the usable and reliable range of results produced by that system (Ezzelle *et al.,* 2008).

The responsibilities of the Laboratory Manager can be divided into five major areas[1]: Principles of Laboratory management; Human resource management; R&D management; Financial (procurement)

1 Source: Laboratory management: Edited by the IE Publications Department. Original version

management and operations. Some Lab Manager's duties include the following:

- Create a healthy, safe, and productive workplace.
- Manage (plan, organize, and direct) the daily work activities of the Laboratory while adhering to organizational policies and goals.
- Communicate thoughts orally and in writing in a clear, well-organized manner.
- Monitor and maintain Laboratory-related documentation, equipment, and supplies.
- Manage scientific and Laboratory practices and procedures by complying with and adhering to national, state, and local standards, policies, protocols, and regulations.

Efficient Laboratory management is also assured when Laboratory Managers build compliance with regulatory issues into the culture of their Laboratories. It is therefore imperative for the Laboratory Manager to be involved in the design, implementation, and oversight of a site-specific, written QC program which defines procedures for monitoring analytic performance and consistent identification, documentation, and resolution of QC issues (Ezzelle *et al.,* 2008). The Laboratory Managers's knowledge of Laboratory or research ethics and on general regulatory requirements and safety guidelines is critical (Hruz, 2008). Some guidelines worth referencing are:

- World Health Organization Laboratory Biosafety Manual (WHO LBM), a compact and convenient directory for practical use (Stavskiy *et al.,* 2003);
- Russia's Sanitary Rules (SRs) which are regulatory sanitary norms that are considered mandatory by any institution regardless of the department it refers to or the form of the property (Stavskiy *et al.,* 2003);
- The Centre for Disease Control and Prevention-National Institutes of Health Biosafety in Microbiological and Biomedical Laboratories (BMBL) of the United States of

America. The BMBL is actually a guideline that provides recommendations for application of safety practices and risk assessment and is of particular importance in providing guidance in the building of new Laboratories or remodeling existing ones.

Other safety guidelines and regulations include HIPPA; Occupational Safety and Health Administration (OSHA) of the U.S. Department of Labor; Radiation Safety; Environmental Safety; Animal Welfare; and IRB (Hruz, 2008). Moreover, the WHO has an international network of collaborating centers for infectious disease diagnostics and research which can serve as a strategic channel to promote responsible Laboratory practices globally. It is recommended that the WHO should require adherence to the WHO Biosafety manual as one of the terms of reference for such collaborations (Gaudioso and Zemlo, 2007). Still with regards to collaboration, Laboratory managers could improve on their competences by outsourcing expertise from professionals like Research and Innovation Management Services bvba (RIMS)[2] which specializes in international research collaborations, valorization and technology transfer, including help with administration and training in international research management.

The Quality Manager

One very important duty post in the Laboratory setting is that of the Quality Manager. In fact, each Laboratory should appoint a Quality Manager who should be familiar with the Laboratory's existing quality system, to co-ordinate all activities related to seeking accreditation. The Quality Manager is a direct report to the Laboratory Manager, though from time to time he may give feedback to the Management of the institution on the overall status of the Laboratory's quality system. Whatever the case, priority reporting of the Quality Manager should be directed to the Laboratory Manager

2 Details on Research & Innovation Management Services bvba (RIMS) can be found at http://www.rimsinternational.com/

and all reports or feedback to the institutions Management must be in consensus or authorized/approved by the Laboratory Manager.

The IT Manager

The Information Technology (IT) Manager is a direct report of the Laboratory Manager. He caters for all issues related to information and data management. He should be a qualified IT professional capable of efficiently managing the Laboratory Information Management System (LIMS). He works in close collaboration with the Documentation Officer.

Documentation Officer

The Laboratory Documentation Officer's role is that of a general secretary who receives, conceives and compiles or archives all documentation pertaining to Laboratory operations such as Standard Operation Procedures, norms, quality manuals, sample information and data sheets etc. He actually monitors the traceability of Laboratory documentation (hard and soft copies) in close collaboration with the IT Manager. He is a direct report to both the Laboratory Manager and the IT Manager.

Financial Officer

The Laboratory Financial Officer is in charge of all monetary transactions regarding Laboratory operations such as transactions relative to procurement of Laboratory equipment and supplies. He keeps records and traceability of all income and expenditure by the Laboratory and reports directly to the Laboratory Manager in conformity with the requirements of the financial department of the institution. The Laboratory Financial Officer should therefore be the respondent of Laboratory financial audits.

Chief Laboratory Technician

The Chief Laboratory Technician (CLT) is actually the operator of routine tasks in the Laboratory, assisted in his duties by the various Laboratory technicians. He has an overview and technical knowhow of the functioning of all appliances and masters all experimental protocols and therefore ensures the quality of Laboratory assays by personally performing them or by guiding the other technicians through the processes. The CLT holds the daily register of the Laboratory in which he records all daily Laboratory operations and pertinent remarks. He also prepares and presents all assay results in the right format to the Laboratory Manager on a regular basis.

Before we close this section, it is worth noting that the proposed organigramme for efficient Laboratory management (Fig. 1) is not exhaustive and should serve as an inspiration for adaptation of a convenient management system to each Laboratory depending on the size and kind of activity undertaken within the Laboratory setting. A very heavy or too scanty administrative system might incidentally influence the quality of work produced by the Laboratory. We have intentionally left our pertinent issues like the Legal department and Procurement Office which are all very important for the smooth running of the Laboratory and which in certain instances could be compensated by such offices at the level of the institution's management. For instance, the organigramme of a small Laboratory like the Lipids Analysis Laboratory of IRAD La Dibamba which I currently manage is limited to a Laboratory Manager and a CLT. All other activities are carried out by technicians under the coordination of team leaders. Meanwhile, a large research facility like the BecA-ILRI Hub in Nairobi, Kenya, which is made up of several Laboratories of various Biosafety levels require a more consistent management team for its smooth running.

Chapter 2

Laboratory design considerations

Objectives and preconditions

The building design of a Laboratory facility is of crucial importance since the outcome of a study may be influenced by the state and condition of the test system at the time of the study. The Laboratory or test facility should be designed safe enough to get the validation results confidently (Isin Akyar, 2011). The primary objective in Laboratory design is to provide a safe environment for Laboratory personnel to conduct their work with maximum flexibility. Specific Laboratory design considerations are included depending on the size of the facility and type of work practices to be undertaken in the lab. Right from the point of Laboratory building design, all health and safety hazards must be anticipated and carefully evaluated so that protective measures can be incorporated into the design (Abad Morejon de Giron *et al.*, 2012; Stanford LSDG, 2013). In addition, carrying out research with dangerous or infectious disease causative agents requires appropriate safety conditions (Stavskiy *et al.*, 2003).

However, no matter how well designed a Laboratory is, education of the facility users is essential since improper usage of its facilities will always defeat the engineered safety features. More specifically, special consideration should be given to the choice of fireproof construction for the building, and selection of the site shall be such as to minimize the risk of landslide or flood damage (Abad Morejon de Giron *et al.*, 2012; Stanford LSDG, 2013). Laboratory building design must therefore be done following consultation of

Regulations, Consensus Standards and References that regulate the putting in place of a facility in a given context. They could involve but may not be limited to the labor code, building code, national electric code, standards for Laboratory ventilation, and standards for thermal environment conditions for human occupancy, radiologic health standards, guidelines for research involving recombinant DNA molecules, guides to reduce risks of nonstructural earthquake damage, etc[3].

For each country, several regulations might be required from various ministries for the establishment of a test facility (Stanford LSDG, 2013). In addition, consideration must be given to Designer Qualifications, Occupancy Classification and Environmental Permits which are basic requirements to be considered in building a Laboratory facility. In this regard, the designer must have the appropriate professional license in his/her area of expertise. Occupancy classification should be based on an assessment of a projected chemical inventory of the building. Prior to the final design, the fire safety organization will need to assign an occupancy official to ensure compliance with the building codes. With regards to environmental permits, project managers must consult with the appropriate administrative services to identify permitting and pollution abatement engineering requirements for the building. This should be done well before key resource allocation decisions are made (Stanford LSDG, 2013).

Safety issues in Laboratory building design

Since the handling and storage of hazardous materials inherently carries a higher risk of exposure and injury, it is important to segregate Laboratory and non-Laboratory activities. The design of the Laboratory building must incorporate adequate additional facilities for food storage and/or consumption and personal hygiene tasks. It is not appropriate to store and consume food, apply cosmetics or lip

3 Source: FEMA, Reducing the Risks of Nonstructural Earthquake Damage

balm in areas they may be contaminated by any toxic material or pathogen. It is thus important to separate office spaces for Laboratory employees such that storage, consumption of food, applying make-up etc. are done far from areas where hazardous materials are used and/ or stored. Therefore, safety will be better assured if non Laboratory work and interaction is conducted in a space separated from the Laboratory (Stanford LSDG, 2013).

Public access to Laboratory personnel in office rooms with separate corridor access is highly desirable. It is recommended that the Laboratory be completely separated from outside areas. Having enclosed Laboratories bound by four walls will help contain spills and keep unauthorized personnel from entering areas where hazardous operations are performed. In addition, the Laboratory should have appropriate means (like lockable doors and lockable cabinets) of securing specifically regulated materials such as controlled substances and select agents and radioactive materials (Stanford LSDG, 2013) thereby ensuring Biosecurity.

The status of windows, sinks, storage, gas lines and waste management systems must equally be compliant. If the Laboratory has windows that open, they must be fitted with insect screens since insects, particularly flies are known to be potential carriers of disease. Sink drain traps shall be transparent and easy to inspect or have drain plugs to facilitate spill control. The Laboratory must contain a sink for hand and face washing located close to the egress. Sink lips should be designed to completely separate the Laboratory bench or fume hood work area from the sink drain.

Chemical storage shelves should not be placed above Laboratory sinks. The storage cabinets will be used based on the chemical inventory and use projection provided by the Principal Investigator or Laboratory Manager. Floors in storage areas for corrosive liquids shall be of liquid tight construction. Materials which in combination with other substances may cause a fire or explosion, or may liberate a flammable or poisonous gas, must be kept separate. When designing the shelves, it is important to provide enough space for secondary containers. Designated storage space should be provided for Laboratory carts whose location must not reduce the width of corridors or aisles to less than code-required widths. The Laboratory carts should be secured with earthquake restraints when not in use.

Auxiliary valves for gas and vacuum lines should be located outside the Laboratory. Flexible connections should be used for connecting gas and other plumbed utilities to any freestanding device such as Biosafety cabinets and incubators. A shutoff valve should be located within sight of the connection and clearly marked. An automatically triggered main gas shutoff valve for the building shall be provided for use in a seismic event. In addition, interior manual shutoff valves shall be provided for both research and teaching areas.

If there are large sections of glass incorporated in the Laboratory building, they shall be shatter resistant. In the event of a severe earthquake, as the glass in cabinets and windows breaks, the shards need to be retained to prevent injury (Stanford LSDG, 2013). Any equipment that has the potential for falling over during an earthquake such as shelves should be permanently braced or anchored to the wall and/or floor. This practice assures that safety while exiting is not compromised.

Safe access

Laboratory shelving should not be installed at heights and distances which require difficult access by workers. Installation of high shelving, above Laboratory benches in particular, can create several potential hazards including spill and exposures to chemicals, radiological or biological agents. A system for ensuring safe access would include prohibition on the materials stored on shelves, limitations on the frequency of use, availability of ladders or ladder stands, training on the use of ladders, etc. Moreover, bench spacing which must be included in specifications and plans shall be considered to provide ease of access or facilitate departure in the event of an emergency. A pathway clearance must be maintained at the face of the access or exit door. The Laboratory doors should be automatically self-closing with a minimum effort required to allow access and egress for physically challenged individuals. Doors which swing in the direction of egress will facilitate occupant departures from Laboratories during emergencies. Laboratory desks should be located near exit ways and in the path of fresh make up air. This will ensure that in the event of an emergency, employees do not have to pass through more hazardous areas to exit the Laboratory. Laboratory benches must not impede emergency access to an exit. This is also applicable to placement of other furniture and appliances such as chairs, stools and refrigerators (Stanford LSDG, 2013).

Sanitation considerations

Laboratories are supposed to be designed such as to ease cleaning. Spaces between benches, cabinets, and equipment must be accessible for cleaning and allow for servicing of equipment. The floor must be non-pervious, one piece, and with covings to the wall and cabinets to ensure that spills cannot penetrate underneath floors or cabinets. Walls are to be painted with washable, hard and non-porous paints. Wooden materials are not appropriate for walls or floors because they can absorb hazardous and/or potentially infectious material, particularly liquids, making decontamination or remediation impossible. Tiles and wooden planks are not appropriate because liquids can seep through the small gaps between them. Also, wood burns rapidly in the event of a fire. All furniture must be sturdy. Bench tops must therefore be a seamless one-piece design to prevent contamination and together with counters must be impervious to the chemicals used. The counter top should incorporate a lip to help prevent run-off onto the floor. Furniture design must therefore comply with basic Design and Construction Standards (Stanford LSDG, 2013).

Electrical and plumbing considerations

In terms of illumination, all Laboratory areas should be provided with adequate natural or artificial illumination to ensure sufficient visibility for operational safety. The room design must take into consideration concerns such as electrical demand prior to occupancy to avoid a potential power failure. The Laboratory should be fitted with an adequate number of electrical outlets, which can accommodate electrical current requirements with an additional 20-40% capacity since the Laboratory may have several equipment which require large amounts of electrical current such as freezers, Biosafety cabinets, centrifuges, etc. Protection to electrical receptacles should be provided. Receptacles that are not readily accessible or receptacles for appliances occupying dedicated space, which are cord-and-plug connected, should be exempted. The circuit breakers for key electrical appliances should be located outside the Laboratory since the Laboratory may be unsafe to enter in the event of an emergency (Stanford LSDG, 2013).

Chapter 3

Good Laboratory Practices

Good Laboratory Practice (GLP) was developed from the Good Manufacturing Practices (GMP) concept that originated in the pharmaceutical industry in the nineteen sixties as a consequence of a series of incidents related to drug manufacturing that ended in severe harm to consumers. GMP established how to implement and perform every process involved in drug manufacturing, with the intended goal of proactively preventing any changes in the identity, safety and efficacy of pharmaceutical products. The GMP established the requirements to ensure the reliability of results from laboratory assays, and these principles later evolved into the current framework known as GLP (Yeniseis, 2008). GLP is a set of rules, operating procedures and the proper practices that ensure the reliability of the data generated by laboratories. GLP is thus a quality system concerned with the organization, process and conditions under which tests are conducted. Basic requirements of Good Laboratory Practices have been outlined by Yeniseis (2008). The principles or parts of Good Laboratory Practices are: organization and personnel, installations and facilities, documentation, equipment and instruments, materials and reagents, reference and assay samples, assay methods' validation, self-inspections and audits, and quality assurance for the assays (Yeniseis, 2008).

GLP Principles include the requirements for the accommodation and citing of test systems, for their maintenance and utilization, and for the associating documentation. It aims at supplying the essential basis for confidence in the results obtained from test systems

(Isin Akyar, 2011). The implementation of GLP helps to promote quality and validity of test data used for determining the safety of products. GLP enhances smooth Laboratory operations, ensures the production and management of high quality data, enhances a stressless environment in the Laboratory, and promotes team spirit and the safety of all Laboratory staff.

GLP is complementary to Quality Management Systems (QMS) which the ISO 9000:2000 standards define as the management system that establishes and controls the relationship of an organization with quality (Yeniseis, 2008). The general requirements of Quality Systems lead to the compliance with the specific requirements of GLP. The common goal of these two concepts is to establish the principles and requirements for an appropriate level of quality assurance. The outcome of GLP complemented with a Quality Management System is resource optimization, efficiency and efficacy in Laboratory operations and management (Yeniseis, 2008).

GLP involves basic housekeeping rules like cleanliness (proper waste management), orderliness and coordinated organisation, use and maintenance of Laboratory equipment and other materials and reagents. An example of a GLP is the use of test items in studies only if they can safely be regarded as being in their pure state (Isin Akyar, 2011). Efficient implementation of GLP involves labelling of equipment, materials and reagents, the establishment and effective use of Standard Operation Procedures (SOPs), effective use of Laboratory Quality Manual (QM) and Material Safety Data Sheets (MSDS) for all materials used within the Laboratory environment.

Use of SOPs and Quality Manual

Controlled documents in the Laboratory should normally include Standard Operating Procedures (SOPs), calibration tables, charts, text books, reference material, software etc. These documents which may be hard copies or electronic documents should be available at their point of use and must not be copied to avoid the possibility of an uncontrolled copy being in use which is no longer valid. Standard

Operating Procedures and controlled documentation may pertain to Quality Assurance, Health and Safety, Environmental Management or general administration and management within the Laboratory setting (FAO, 2013). Written SOPs are an important tool to assure that recurring tasks in a Laboratory are performed in a consistent manner. Safety elements are critical components of the document (Ingrid Schmid, 2012). The laboratory must write SOPs in a manner and language that is appropriate to the laboratory personnel conducting the procedures. Such SOPs must be readily available in the work areas and accessible to testing personnel for all laboratory activities to ensure the consistency, quality, and integrity of the generated data (Ezzelle *et al.*, 2008).

A Quality Manual (QM) is a documentation process that includes procedures and associated records for Institutional references. Laboratory Managers should consult with existing programmes or institutions that could assist in the development and implementation of SOPs and Quality Manuals for smooth Laboratory operations. For instance, several Public Employees Occupational Safety and Health (PEOSH) Standards require the development and implementation of written programs or plans to establish standard operating procedures designed to protect employee safety and health. In order to assist employers in complying with these requirements, the New Jersey Department of Health and Senior Services (NJDHSS) - PEOSH Program, has developed model programs for several standards. These model programs can be obtained from the Public Employees Occupational Safety and Health (PEOSH) website www.state.nj.us/health/eoh/peoshweb. Furthermore, in the USA, free on-site safety and health consultation services are available in all states to employers who want help in establishing and maintaining a safe and healthful workplace.

Moreover, the Occupational Safety and Health Administration (OSHA) of the U.S. Department of Labor Consultation Service delivered by state governments employ professional safety and health consultants who offer service assistance such as appraisal of all mechanical systems, physical work practices, occupational safety and health hazards of the workplace, and all aspects of the employer's present job safety and health program. In addition, the service offers assistance to employers in developing and implementing an effective workplace safety and health program that corrects and continuously addresses safety and health concerns (OSHA, 2000). OSHA standards, interpretations, directives, and additional information are available on the Internet at http://www.osha.gov/, meanwhile a wide variety of OSHA materials can be purchased from the Government Printing Office (OSHA, 2000).

Use of Material Safety Data Sheets

One of the most important and indispensable safety tools for Laboratory operators is the Material Safety Data Sheet (MSDS). This is a document which provides guidance for handling a hazardous or non hazardous substance and information on its composition and properties. MSDS is also called Safety Data Sheet (SDS), or Product Safety Data Sheet (PSDS). It is an important component of product stewardship and workplace safety. They are a widely used system for cataloging information on chemical compounds, and chemical mixtures. In general, MSDS are not intended for use by the general public. They mainly focus on the hazards of working with the material in an occupational setting like the Laboratory. The Laboratory must therefore have Material Safety Data Sheets (MSDS) or equivalent for each hazardous material/chemical used (Ezzelle *et al.,* 2008).

MSDS is intended to provide Laboratory workers and emergency personnel with procedures for handling or working with substances in a safe manner, and includes information such as physical data (melting point, boiling point, flash point, etc.), toxicity, health effects, first aid, reactivity, storage, disposal, protective equipment, and spill-handling procedures. MSDS formats can vary from source to source within a country depending on national requirements. Most Safety Data Sheets can be downloaded from the internet free of any charges by simply hitting "MSDS" and the name or keywords relative to the material in question in any search engine.

Labeling as a GLP

The simplest way of addressing potential risks or hazards at the workplace is by labeling. The Laboratory management must ensure that all materials or substances and reagents are properly labeled on the basis of physico-chemical, health and/or environmental risk factors. The labeling must include information or codes describing content, safety signs, storage requirements, date opened, prepared, or reconstituted by the laboratory, and the initials of personnel who installed, prepared or reconstituted the materials and reagents, and expiration date (Ezzelle *et al.,* 2008). Most Laboratory materials come with simple hazard labels on them. Labels may include hazard symbols such as the European Union standard black diagonal cross on an orange background, used to denote a harmful substance.

Working in the Laboratory with unlabeled materials is a highly risky behaviour. For instance, testing by clinical personnel at the point of care is more prone to errors than analyses conducted by experienced Laboratory professionals with training in error recognition and prevention. Hence, operators in a clinical Laboratory can inadvertently omit or misidentify a patient and associate the labeling of his/her sample with the wrong patient. In such instances, the optimum Laboratory control processes will be worthless if the specimen is mislabeled with another patient's identification (Nichols, 2011).

Work practice and engineering controls

Engineering controls as GLP involve all endeavors to physically manipulate an appliance or the working environment to eliminate a hazard or prevent the exposure of operators to it. It is obvious that no single control process can cover all devices and types of errors in a Laboratory since Laboratory devices differ in design, technology, function and intended use (Nichols, 2011). Some devices have internal checks which are performed automatically with every specimen, while the possibility of other errors is reduced through engineering into the device by the manufacturer (Nichols, 2011). After identifying a potential hazard, engineering control could be implemented simply by adding a guard to the machine or building a barrier between employees and the hazard. Meanwhile work practice control could be implemented by simply removing employees from exposure to potential hazard or by changing their job description or position (OSHA, 2000).

Use of Personal Protective Equipment

The use of personal protective equipment (PPE) in a Laboratory setting is consistent with any workplace safety requirements. Laboratory safety is ensured when all Laboratory users are provided with protective equipment that ensures a safe and healthful work environment free from recognized hazards. Personal protection refers to the placement of physical, primary barriers between operators and hazards as a means to protect the operators from exposure in the event of a breakdown of mechanical barriers or engineering devices (Ingrid Schmid, 2012). It is essential that employers protect their employees from workplace hazards such as machines, work procedures, and hazardous substances that can cause injury. The preferred way to do this is through engineering controls or work

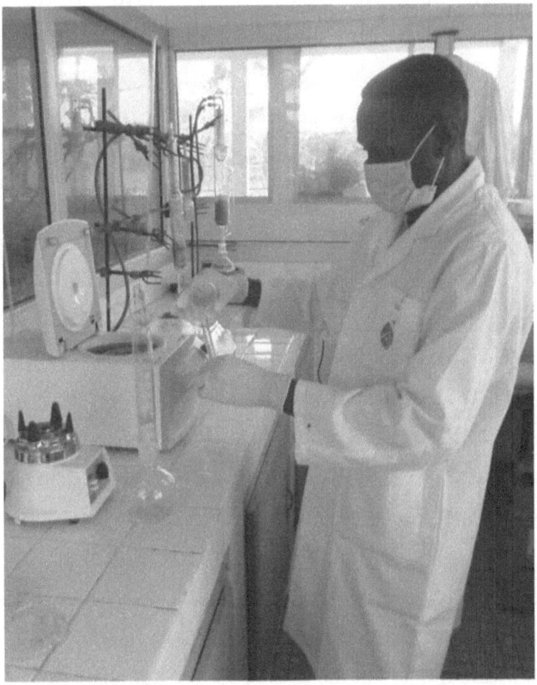

practice and administrative controls; but when these controls are not feasible or do not provide sufficient protection, an alternative or supplementary method of protection is to provide workers with Personal Protective Equipment and the know-how to use them properly (OSHA, 2000).

The extent and types of PPE used differ between Laboratories. Hence, Laboratory management should select appropriate PPE to suit the kind of activities and degree of risk exposure. For instance, BSL-III and BSL-IV practices require wearing of respiratory protective devices and equipment that protect against aerosol exposure (Ingrid Schmid, 2012).

As mentioned earlier, critical aspects in the use of PPE involve placing physical barriers between the operator and the hazard and mandatory operator training in the SOPs established in the Laboratory. It is recommended to prevent access to the Laboratory by persons not wearing the required personal protective equipment; meanwhile the PPE should always be removed whenever the operator leaves the Laboratory. In addition to the use of PPE, immunization or prophylactic vaccination against infectious agents should be offered to personnel (Ingrid Schmid, 2012). Moreover, post-exposure prophylaxis should be available in any Laboratory involved in the manipulation of hazardous substances or sorting of infectious specimens (Wang *et al.,* 2000; Mikulich and Schriger, 2002; Schriger and Mikulich, 2002; Ingrid Schmid, 2012).

PPE should be used in conjunction with these controls to provide for employee safety and health in the workplace. The basic element of any management program for PPE is an in-depth evaluation of the equipment needed to protect against the identified or potential hazards at the workplace. Management dedicated to the safety and health of employees should use such an evaluation to set standard operating procedures for personnel. Furthermore, the employer must provide training, as mentioned earlier, to each employee required to use PPE. Training will include: when PPE is necessary, what PPE is necessary, how to properly wear and take off PPE, how to properly inspect PPE for wear or damage; the limitations of the PPE, and the proper care, maintenance, useful life, and storage or disposal of the PPE. The Laboratory Manager has to certify in writing that the staff has received and understands the training (Clifton *et al.,* 2003). The PPE must be used and maintained in a sanitary and reliable condition. In fact, PPE should not be used as a substitute for engineering, work practice, and/or administrative controls.

Establishing a PPE program

A PPE program is actually a framework that sets out procedures for selecting, providing, and using PPE as part of routine operations in a setting. A written PPE program is easier to establish and maintain as company policy and easier to evaluate than one which is not written. The following checklist can be used to establish a PPE program for a test facility or company (OSHA, 2000):

- Identify steps taken to assess potential hazards in every employee's work space and in workplace operating procedures.
- Identify appropriate PPE selection criteria.
- Identify how to assess employee understanding of PPE training.
- Determine how to enforce proper PPE use.
- Determine what to provide as required medical examinations.
- Identify how and when to evaluate the PPE program.
- Identify how to train employees on the use of PPE.

The employer must assess the workplace to determine if hazards which necessitate the use of PPE are present, or are likely to be present. Such a workplace survey to identify potential hazards could be by observing the environment in which the employees work, asking employees how they perform their tasks and looking for sources of potential injury or hazards such as (OSHA, 2000): objects that might fall from above; exposed pipes or beams at work level; exposed liquid chemicals; sources of heat, intense light, noise, or dust; equipment or materials that could produce flying particles etc. If such hazards are present, or are likely to be present, the employer shall select, and have each affected employee use the types of PPE that will protect against the identified hazards. The employer shall verify the hazard assessment in writing and make sure that the PPE properly fits each affected employee (Clifton *et al.,* 2003).

Categories of PPE

PPE includes all clothing and other work accessories designed to create a barrier against workplace hazards such as devices for protecting the eyes, face, head and extremities. In general, PPE includes such items as goggles, face shields, safety glasses, hard hats, safety shoes, gloves, vests, earplugs, and earmuffs. Protective shields, respirators and rubber insulating equipment like gloves, sleeves, and blankets are also considered as PPE (OSHA, 2000; Clifton *et al.,* 2003). PPE can be grouped into the following categories (Clifton *et al.,* 2003):

Body protection: Employees whose bodies are exposed to irritating dust or chemical splashes, sharp or rough surfaces, extreme heat, acids or other hazardous substances (OSHA, 2000) are supposed to use specific PPE.

Eye and face protection: Appropriate eye or face protection should be used by employees when exposed to flying particles, potentially hazardous chemical gases or vapors, liquid chemicals, potentially injurious light radiation and the like (OSHA, 2000; Clifton *et al.,* 2003). Where there is a potential for exposure to corrosive materials, suitable facilities for flushing the eyes and body such as emergency eye washes and showers must be available in the Laboratory for emergency use (Clifton *et al.,* 2003).

Ear protection must be used by employees exposed to loud noise from machines or tools (OSHA, 2000).

Respiratory protection: It is recommended that the employer provides respirators when such equipment is necessary to protect the health of the employees. Such equipment has to be applicable and suitable for the purposes intended (Clifton *et al.,* 2003; HSE, 2012). Some circumstances that require the use of respirators include (Clifton *et al.,* 2003): where exposure levels exceed the permissible exposure limit (PEL); during maintenance and repair activities especially where exposures exceed the PEL and where engineering or work practice controls are not feasible or are not required such as in regulated areas, or where the feasible engineering and work practice controls implemented are not sufficient to reduce exposure to or below the PEL, and in emergencies.

Head protection: Head protection of employees is assured by the use of protective helmets including those designed to reduce electric shock hazard (from electrical conductors for instance) where necessary. Helmets must be worn when working in areas where there is a potential for injury to the head from falling objects (OSHA, 2000; Clifton *et al.,* 2003).

Electrical protective equipment: These equipment are very important and should comply with the design requirements of existing standards (Clifton *et al.,* 2003).

Hand and feet protection: Such equipment are necessary for employees whose hands and feet are exposed to hazards such as those from skin absorption of harmful substances; severe cuts or lacerations; punctures; chemical burns; thermal burns and harmful temperature extremes. The selection of appropriate hand protection should be based on evaluation of the performance characteristics of the hand protection relative to the task to be performed, conditions present, duration of use and the hazards and potential hazards identified (OSHA, 2000; Clifton *et al.,* 2003).

The use of PPE to protect employees against hazards does not eliminate the fact that Management should institute feasible engineering, work practice, and administrative controls to eliminate or reduce the hazards (OSHA, 2000). Management should equally check whether there are ways other than PPE in which the risks can be adequately controlled (HSE, 1992). It is worth of emphasis for Management to ensure that PPE which offers adequate protection for its intended use is provided and that those using it are adequately trained in its safe use. Care should also be taken on whether the PPE provided is properly maintained. Any defects reported should be corrected on time and the PPE should always be returned to its proper storage after use.

Chapter 4

Biosafety and Biosecurity

Many countries in the world have chosen to seek economic growth through large national investments in genetic engineering and biotechnological research (Gaudioso and Zemlo, 2007). Such investments have to be guided by sufficient planning toward securing and maintaining adequate Biosafety and Biosecurity precautions (Gaudioso and Zemlo, 2007). Biological safety requires the strengthening of the regulatory requirements and operational and safety programs (Tun *et al.,* 2007). Safety policies (e.g. Standard Precautions/Universal Precautions Policy, Chemical Hygiene/ Hazard Communication Plan, Waste Management Policy, Safety Equipment, and general safety policies) defined according to regulatory organizations such as the OSHA or ISO must be present in the Laboratory (Ezzelle *et al.,* 2008). More detailed technical guidance is to be provided to Biosafety Officers and Laboratory Managers (Gaudioso *et al.,* 2006). Since Biosafety and Biosecurity jointly define the Laboratory operating environment (Gaudioso *et al.,* 2006), it is critical to carefully plan the implementation of Biosecurity to avoid conflicts with Biosafety.

The rationale for Biosafety and Biosecurity

The World Health Organization (WHO) recognizes Biosafety and Biosecurity as important international issues through a manual which encourages countries to prepare specific codes of practices for the safe handling of potentially hazardous agents (Tun *et al.,*

2007). According to FAO, even the use of veterinary medicines should be in line with established principles on prudent use to safeguard public and animal health (Bondad-Reantaso *et al.,* 2012). The same stipulates that the use of such medicines should be part of national and on-farm Biosecurity plans and in accordance with an overall national policy for sustainable aquaculture. Biosafety and Biosecurity must therefore take a central stage position in the present climate where emerging and re-emerging infectious diseases and threats of bioterrorism and biological attacks are of global concern (Tun *et al.,* 2007). Their implementation is thus strategic in the global battle against diseases such as Avian Influenza, Severe Acute Respiratory Syndrome virus (SARS), Nipah, Chikungunya, epidemic meningitis, hantavirus, Human Immunodeficiency Virus (HIV), bovine spongiform encephalopathy (BSE) [BSE which also affects humans is the commonly called mad cow disease which first came to the attention of the agricultural and scientific community in 1986 with the appearance of a new form of neurological disease in cattle in the United Kingdom (Leunda *et al.,* 2013)], and Rift Valley Fever, among many others including bioterrorism considered as a "deliberately reemerging infectious disease" (DRID) (Morens *et al.,* 2004; Gaudioso and Zemlo, 2007).

Furthermore, with the first DNA experiments (Cohen *et al.,* 1972), concern arose in the scientific community regarding the possible hazards associated with the recombinant DNA technologies (Cardoso *et al.,* 2005) leading to the holding of the Asilomar Conference in California [1975] during which rules for the control of these risks were established with ethical criteria and operational procedures necessary to ensure that any new experiments in the field of genetic engineering were ethical and safe (Berg *et al.,* 1975; Cardoso *et al.,* 2005). The Asilomar conference was also the beginning of a series of polemics and discussions regarding the potential risks and benefits these techniques could bring to the development of new species (Gura, 1999; Cardoso *et al.,* 2005).

Joint international contingency programs and the need for biological safety and security considerations are thus needed to counteract the aforementioned existing threats especially with regards to the use of bioagents by terrorists (Stavskiy *et al.*, 2003; Bakanidze *et al.*, 2010; Xin Pan, 2012). The increased recognition of bioterrorism threat has expanded the realm of Biosafety to merge it with Biosecurity; making Biosafety to be viewed in a new broader sense. Biosafety is considered from this new perspective with respect to two issues pertinent to bioterrorism and Biosecurity. These issues are: airborne exposure limits (AEL) for biological threat agents, biowarfare or BW agents; and the sensitivity limits for the real-time biosensors that should detect BW agents as well as the significance of these limits for Biosafety and Biosecurity (Sabelnikov *et al.*, 2006). Emergency Biosafety response strategies or guidelines to deal with any kind of terrorist event should incorporate several factors such as immediate mitigation, organizational, and postevent procedures like clean up and decontamination. These responses should be based on scientifically proven exposure and health risk assessment for the materials involved (chemical, biological, or nuclear/radiological); effective detection, identification, and decontamination technologies; and accepted clean-up standards (Sabelnikov *et al.*, 2006) which are all possible through consistent Laboratory analysis and practices.

Risk assessment for Biosafety and Biosecurity

A genuine scientific Laboratory should have in place a risk assessment and risk management plan. The employer must assess the workplace to determine if hazards are likely to be present (Ezzelle *et al.*, 2008). The consensus among European *BSL-IV* experts is that, in the interest of safety, standard practice should be for all Laboratories to perform a risk assessment before any activity is undertaken (Ippolito *et al.*, 2009). Thus, risk or probability of hazard is the basis of Biosafety (Cardoso *et al.*, 2005). Before 1975, little was known about procedures for the assessment and prevention of hazards, and Biosafety was restricted to information about risks inherent in scientific research (ABSA, 2001; Cardoso *et al.*, 2005). The historical milestones in the process of risk assessment in activities involving biological agents were established by studies carried out by Meyer and Eddie in 1941, followed by those of Sulkin and Pike in 1949 (Cardoso *et al.,* 2005). These studies resulted in the first containment and control procedures that paved the way towards the consideration of Biosafety as a discipline (Cardoso *et al.,* 2005). By 1980, the American Biological Safety Association (ABSA) was founded as the first scientific association in the world established to develop Biosafety as a scientific discipline (Cardoso *et al.,* 2005). ABSA provides scientific interchange in the field of risk assessment and risk management procedures, builds human resources in the field, and disseminates scientific information through publications in different areas of Biosafety (ABSA, 2003).

Laboratory activities are supposed to be subject to regulations on the protection of workers from risks related to exposure to and on the contained use of biological/hazardous agents at work (Leunda *et al.,* 2013). According to the European Union Directive 2009/41/EC, a contained use of genetically modified microorganism (GMM) and/or pathogenic microorganism should be subject to a risk assessment in order to define proper risk management including the adoption of adequate containment measures and work practices (Leunda *et al.,* 2013). Laboratory risk assessment therefore helps to provide a high level of safety for the general population and the environment (Leunda *et al.,* 2013). The risk assessment is based on the identification of potentially harmful properties of the biological or hazardous agents such as pathogenicity, transmission mode, persistence and stability of the agent in the environment and availability of effective prophylaxis or therapy (Leunda *et al.,* 2013). A facility risk assessment is therefore the foundation of an efficient Biosecurity program. Risk assessment and the enhancement of safety could be facilitated by Laboratories creating process maps that outline all the steps involved in the different testing processes underway in the Laboratory (Nichols, 2011). All Laboratories must have a security plan designed according to a site-specific risk assessment and must provide graded protection in accordance with the risk of select agents or toxins, given their intended use (Gaudioso *et al.,* 2006). Risk assessment could equally be performed before Laboratory dismantlement, taking into account the present and future activities to be performed in the Laboratory

(Leunda *et al.*, 2011; Leunda *et al.*, 2013). Medical Laboratories may use the ISO 15190 as a guideline for preparing a written Laboratory safety procedure, training staff on techniques of safety precaution, e.g. not to drink or eat in the Laboratory, wear personnel protective equipment like gloves, masks and gowns when working with infectious material etc. Moreover, the Laboratory should perform testing in a Biosafety cabinet class 2 or fume hood, where necessary (Panadda, 2007).

Laboratory Biosecurity

It can be inferred from the preceding sections that Laboratory Biosecurity encompasses physical and administrative measures designed to protect biological material like pathogens, toxins, and sensitive information that could cause harm to health or economic loss resulting from malicious release, intentional loss, theft, sabotage, or misappropriation, and subsequent misuse (Summermatter, 2009). Biosecurity is a part of Biosafety and the two are components of Good Laboratory Practice (GLP). Biosecurity is important in the sense that it actually improves security in the Laboratory, increases emergency preparedness, protects the reputation or image of the institution, mitigates liability exposure and protects public health, employees and research in general. Biosecurity may be assured by ensuring controlled access to Laboratory facility and information, taking into consideration regulations (such as IATA/WHO regulations) for packaging, transfer or shipping of research or biological materials between Laboratories. **Figure 2** indicates some recommended containers used in the transfer of biological materials.

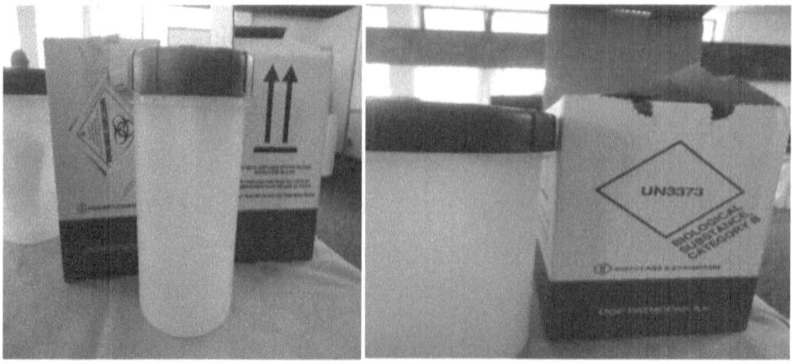

Figure 2. Some recommended shipment containers for biological substances

Note standard UN number (UN3373) and appropriate safety labels. The sample should be put in the screw capped plastic container before it is placed in the cartoon as recommended

Laboratory Biosecurity practices or measures to protect biological research materials must coexist with Biosafety (Gaudioso *et al.,* 2006). Laboratory Managers are responsible for implementing and overseeing the Laboratory's Biosecurity program. Biosecurity program management responsibilities might include identifying the protection objectives, designing the security system, writing security and emergency response plans, conducting regular training and internal reviews, and allocating resources (Gaudioso *et al.,* 2006). However, there are shortcomings for efficient Biosecurity implementation which include the time and effort required by the staff to comply with the regulations and the obvious inconveniences of increased security in the Laboratory setting. Biosecurity measures that are standard Laboratory practices may have greater community acceptance. Efficient implementation of Biosecurity requires time for adaptation, especially in Laboratory settings where a culture of free and open access exists (Gaudioso *et al.,* 2006). It is also important to involve safety officials, law enforcement officers, and researchers into the process of Biosecurity plan development (Gaudioso *et al.,* 2006).

Brief on the American Biosecurity experience

The United States of America (U.S.A.) has enhanced its regulatory approach for security of pathogens and toxins through the passage of the USA PATRIOT Act of 2001 and the Bioterrorism Preparedness Act of 2002. In February 2003, three interim Codes of Federal Regulations (CFRs) became effective which specified security measures for approximately 80 pathogens and toxins known as select agents, deemed to pose a threat to human, animal, or plant populations. The regulations require any Laboratory that possesses one of these agents or toxins to enforce and adhere to specific security measures, which include facility registration; designation of a Responsible Official; security risk assessments for individuals with access (or possession) to the listed agents; Laboratory Biosecurity plans; agent transfer rules; safety and security training and inspections; notification after theft, loss, or release of a listed agent; record maintenance; and restrictions of certain experiments. Other CFRs rules that define the regulatory requirements for Laboratory Biosecurity in the U.S. replaced the interim regulations in March 2005, with a few changes that impact the implementation of Biosecurity in Laboratories. It is worth noting however that facilities that do not work with select agents (SA) are not required to implement these rules (Gaudioso *et al.,* 2006).

Laboratory Biosafety

Biosafety potentials for consideration as a separate academic discipline

From the onset, the main concern of Biosafety was primarily Laboratory safety. Research on Biosafety and related topics is making it a more exact and defined discipline with quantitative and prediction power. The discipline may include areas such as risk assessment and analysis, sampling methods to detect contaminants, analytical method development and validation, and novel methods developed for decontamination (Sabelnikov *et al.,* 2006). With reference to a publication of Centers for Disease Control (USA) (CDC and NIH, 1984), Biosafety is a set of procedures, practices, and equipment for minimizing or preventing hazards posed by pathogens (Cardoso *et al.,* 2005). Biosafety is also defined as "science directed to control and prevent hazards posed by the use of different technologies, be it at the Laboratory, be it applied to the environment," with the basic purposes of ensuring the advancement of technological processes and of protecting human and animal health and the environment (Cardoso *et al.,* 2005).

Biosafety is recognized as a multidisciplinary science, though CDC and NIH (1999) and WHO (2003) focus mainly on biological hazards, considering other hazards posed by the Laboratory work process only as byproducts (Cardoso *et al.,* 2005). According to Summermatter (2009), Laboratory Biosafety is the reduction and prevention of the risk of exposure of workers and the environment to dangerous pathogens. Risk which sets the premise for Biosafety is a part of all scientific activity (UNEP, 1996; WHO, 2003; Cardoso *et al.,* 2005). The basic principle of Biosafety is thus risk control and risk management (Cardoso *et al.,* 2005). The preventive methods used to protect the researcher, the object under research, and the environment constitute the first step in minimizing risk and defines the field of action of Biosafety (Oda and Ávila, 1998; Cardoso *et al.,* 2005). Therefore, Biosafety involves all endeavors aimed at reducing or eliminating accidental exposure of individuals and the environment to potentially hazardous biological agents. Assessing the exposure risk to personnel working in Laboratories or environments where microorganisms are present is only one of the fields of action of Biosafety as a scientific discipline (Cardoso *et al.,* 2005). Biosafety is poised to becoming an obligatory discipline in all scientific graduate courses as is the case at the Oswaldo Cruz Institute in Brazil (Cardoso *et al.,* 2005). Given its global importance and the potential of efficiently and proactively mitigating safety issues, I strongly reiterate this recommendation here that Biosafety as a discipline effectively becomes part and parcel of all scientific graduate courses worldwide. Benchmarking from success stories could be an efficient start point for the implementation of this recommendation, while global regulatory organizations could require this as part of curriculum development and application in higher institutes of learning around the world.

Laboratory classification into Biosafety Levels

Properties of biological test systems are generally more complex and mutable than those of physical and/or chemical test systems. Hence biological test systems need very careful characterization in order to guarantee the quality and integrity of the data derived from them (Isin Akyar, 2011). The European Union legislation considers the classification of workplaces in hazardous zones as compulsory, in order to prevent explosions. The classification procedure requires knowledge about the sources of ignition and the potential sources of risk as described in the ATEX 137 Workplace Directive 1999/92/EC (Di Benedetto *et al.,)* Furthermore, CDC (1974) has presented the criteria for biological risk assessment and the procedures for containing hazards, classifying them into four basic levels (Cardoso *et al.,* 2005). The descriptions for Biosafety Levels I to IV (BSL I to BSL IV) as a form of risk management with recombinant DNA are consistent with the original classification criteria for conventional etiological agents (NIH, 1994; WHO, 2003).

In fact, the classification of Laboratories into various levels of Biosafety is a sure way of creating a safe Laboratory environment. This helps to deal with infectious and pathogenic organisms that may lead to the occurrence of outbreaks due to their rapid spread (Devendra *et al.,* 2014), Laboratory classification on this basis helps to protect Laboratory workers, the environment and community as a whole (Devendra *et al.,* 2014). It is relevant for each Laboratory to harmonize national and international biological safety guidelines regulating the organization and work with hazard group I to IV pathogens in different countries (Stavskiy *et al.,* 2003). A simplified presentation of the four Biosafety levels of Laboratories is as follows[4]:

4 Source: Lab Classification 27/10/05

BSL-I Laboratory

Biosafety Level I (BSL-I) Laboratories are used for the manipulation of microorganisms with low hazard levels. The organisms used here are not known to cause disease in healthy adults (Stavskiy *et al.,* 2003). They are organisms which are classified in risk group 1 such as inactivated or fixed specimens. Biosafety Level I (BSL-I) Laboratories are appropriate for student and undergraduate teaching Laboratories. Laboratory personnel can be protected by standard Laboratory practices in the BSL-I Laboratory setting. Given the low level of risk, work in BSL-I Laboratories may be carried out on the open bench.

BSL-II Laboratory

The Biosafety Level II (BSL-II) Laboratory is appropriate for clinical, diagnostic, industrial, teaching and other premises where work is carried out with microorganisms or material likely to contain microorganisms. Work in such Laboratories involves manipulation of risk group 2 microorganisms which may be associated with human, animal, or plant diseases of moderate severity (Stavskiy *et al.,* 2003). With good microbiological techniques, work with these organisms may be carried out on the open bench. If there is a significant risk of generating aerosols, a biological safety cabinet must be used in the BSL-II Laboratory.

BSL-III Laboratory

The BSL-III Laboratory is appropriate for clinical and diagnostic work with indigenous or exotic microorganisms classified in risk group 3 and considered to present the risk of serious infection of humans, animals, or plants (Stavskiy *et al.,* 2003). An example is the TSE (transmissible spongiform encephalopathy) causing agents classified as risk group III organisms for humans and animals by most of international pathogenic organism classifications (Leunda *et al.,* 2013). In a Laboratory manipulating such a pathogen, protection of public health and environment requires a special attention to decontamination procedures and waste management (Leunda *et al.*, 2013). A BSL-III Laboratory facility provides safeguards to minimize the risk of infection to individuals, the community and the environment.

BSL-IV Laboratory

A Biosafety Level IV (BSL-IV) Laboratory is appropriate for work with dangerous microorganisms that pose a high risk of life threatening diseases and may be spread to the community. They belong to risk group IV microorganisms (Stavskiy *et al.*, 2003). A BSL-IV Laboratory is a facility situated in a building separate from other Laboratories, or constructed as an isolated area within a building. The facility in BSL-IV must include secondary barriers such as: sealable openings into the Laboratory; airlocks or liquid disinfection barriers; a clothing change and shower room contiguous to the Laboratory; ventilation system contiguous to the Laboratory ventilation system; and exhaust air and liquid waste decontamination systems. Biosafety and Biosecurity are the major concerns for use of a *BSL-IV* Laboratory (Ippolito *et al.*, 2009). It is recommended that more than 2 persons should be present at all times while biological work is undertaken in a *BSL-IV* Laboratory (Ippolito *et al.*, 2009; Le Duc, 2009). However, safety and security would be better assured in some situations by video monitoring systems rather than by the presence of a fellow scientist (Ippolito *et al.*, 2009). Preliminary risk assessment is necessary to determine procedures to be used, including whether 2 persons should work together as part of Laboratory procedure. The best approach is to have flexible rules that are objectively assessed according to Laboratory-specific circumstances (Ippolito *et al.*, 2009). Surveillance video monitoring and data storing have their place in protecting Laboratory facilities from unauthorized access and theft of materials, but their effectiveness for ensuring proper handling of pathogens is quite limited. Both Biosafety and Biosecurity must be founded on careful selection and monitoring of staff, without which even the most sophisticated of control systems would fail (Ippolito *et al.*, 2009).

From the above classification of Laboratories into Biosafety Levels, it is worth noting that the context and level of Laboratory facility with its practices and equipment actually determine the classification of a Laboratory in one of the four categories or Biosafety Levels (BSL). The risk group classification of hazardous agents does vary from country to country, even though there are global standard Laboratory practices with many aspects of Laboratory culture shared throughout the world (Tun *et al.,* 2007). For instance, guidelines for Biosafety and Biosecurity of the Ministry of Health of Singapore include regulations regarding the import, transport, transfer, handling, and disposal of human pathogens and their risk group classifications (Tun *et al.,* 2007). An agent classified into Risk Group II in one country may be classified as Risk Group III in another. Dengue Virus Type 1-4 was classified as Risk Group II in Singapore (previously), Canada and Australia but was classified as Risk Group III in Belgium and the European Union (Tun *et al.,* 2006; Tun *et al.,* 2007).

Under the Biological Agents and Toxins Act (BATA) in Singapore for instance, biological agents and toxins are classified into schedules and not Biosafety levels. The BATA differentiates between higher risk group and lower risk group Biological Agents, and also those with potential to be weaponized (Tun *et al.,* 2007). Five schedules in BATA cover a wide spectrum of biological agents and toxins and different levels of controls have been adopted for each schedule, with corresponding risk groups, facility requirements and biological agents in each schedule (Tun *et al.,* 2007).

Laboratory classification based on context and threat

The SARS-CoV experience

The SARS experience can be cited as an example of a pathogen that could be manipulated based on the context and threat it poses. The outbreak of SARS-CoV (Severe Acute Respiratory Syndrome Corona Virus) raised Biosafety concerns and, specifically, the question of risk assessment regarding the contained use of microbes for Laboratory work (Drosten *et al.,* 2003; Ksiazek *et al.,* 2003;

Peiris *et al.,* 2003; Herman *et al.,* 2004). In fact, at the end of 2002, an outbreak of SARS-CoV occurred in the Guangdong Province in China, with the earliest known cases identified in mid-November 2002. The syndrome spread to Asia, North America, Europe, and Africa, with 8,422 suspected and probable cases of SARS including 916 deaths reported to the World Health Organization (WHO) by August 7, 2003 (Herman *et al.,* 2004). In response to the SARS outbreak, the WHO coordinated an international collaboration that included clinical, epidemiologic, and Laboratory investigations, and initiated efforts to control the spread of SARS.

Based on a scientific risk assessment related to the contained use of biological agents, the SARS-CoV was classified as a Risk Group III agent (Herman *et al.,* 2004). In relation to the reported Biosafety assessment, the SARS-CoV could be handled in appropriate Biosafety containment levels to avoid Laboratory-acquired infections and spread of the disease in the human population and the environment. Diagnostic activities with inactivated clinical specimens associated with SARS cases and with specimens originating in countries where SARS is documented but not associated with SARS cases could be performed under Biosafety Level II (BSL-II) conditions. Meanwhile, diagnostic activities involving non inactivated clinical specimens associated with SARS could be carried out under BSL-II containment with BSL-III safety equipment and work practices. Culture of SARS-CoV and all research activities involving SARS-CoV should require BSL-III containment (Herman *et al.,* 2004). There are also guidelines related to the handling of SARS-CoV in clinical, research, or production Laboratories (Herman *et al.,* 2004).

Laboratory-acquired infections of contagious diseases have demonstrated the potential to spread beyond the Laboratory into the community (Gaudioso and Zemlo, 2007). For instance, multiple incidents of Laboratory-acquired SARS (Singapore in September 2003, Taiwan in December 2003, and Beijing in March 2004) occurred after the virus stopped circulating naturally. The Beijing Laboratory acquired infection (LAI) spread into the community, resulting in nine cases of SARS. LAIs occur throughout the world. Examples include Ebola in Russia (2004), vaccinia in Brazil (2002), and tularemia in Boston (2004). However, at least in part because the SARS LAIs occurred after the virus was no longer circulating in humans, LAIs have helped to generate a greater awareness of the need to ensure that dangerous pathogens are handled responsibly (Gaudioso and Zemlo, 2007).

Management of Laboratory waste as a GLP

Historical background of waste management

Waste or refuse management involves initial handling, collecting, transporting, disposing and monitoring of waste materials (Tabasi and Marthandan, 2013). Refuse has not always been a problem for human societies. Nomadic, rural, and agrarian economies have successfully relied upon natural forces to manage their waste. Sunlight and microbes in water and soils readily decomposed garbage and human waste, as long as population densities remained low. When humans were primarily hunters and gatherers, garbage could simply be left where it fell. When people established permanent settlements, the situation began to change. Settlements grew larger, laying the groundwork for the world's first cities and the solid waste problems that accompanied civilization (Phillips, 1998). Most people dealt with solid waste by dumping it somewhere on the floor, in the street, or into decidedly unsanitary open pits (Phillips, 1998). Burning and recycling have also been practiced since prehistoric times. However, these early waste management practices did not have a major impact on the environment. As cities grew, major sanitation problems developed such as frequent and often devastating epidemics. From thence, the primary goal of those who manage solid waste has been to safeguard the public health and with time, many poor solid waste management methods have been eliminated (Phillips, 1998).

Today, with better technologies and a deeper understanding of how to prevent environmental pollution, waste management and disposal have an additional goal: to wisely manage material and energy resources. Unlike earlier goals, this objective cannot be met by waste management professionals alone. This is one of the reasons why in modern times, most nations or institutions are committed to safer, more environmentally responsible management of waste (Phillips, 1998).

In many developed and developing countries, collection, transportation, treatment and disposal of waste are the major challenges for government, organizations and other institutions. Different types of solid wastes depending on the generation resource can be classified into household or municipal waste, industrial waste, hazardous waste and biomedical/Laboratory waste (Tabasi and Marthandan, 2013). In the field of medicine, concern over clinical waste has increased throughout the world. Improper management of clinical waste poses a public health risk, prompting the need for appropriate Clinical Waste Management (CWM) as a crucial issue for maintaining human and public health (Nema *et al.,* 2011; Tabasi and Marthandan, 2013). The CWM practices cover all processes from the point of identification of the wastes, to the place it is disposed in an incinerator (Tabasi and Marthandan, 2013).

Laboratory waste management

Proper management of Laboratory waste is a capital element in efficient Laboratory management. Wastes from a Laboratory are proportionate to the Laboratory size and kind of activities carried out in there. Developing appropriate models for accurate prediction of waste generation rate can be useful in the storage, transportation, treatment and disposal of laboratory waste (Tabasi and Marthandan, 2013). The Laboratory should be designed to safely accommodate the temporary storage of biological, radiological, and non-waste and waste chemicals. It is appropriate and recommended practice for wastes to be stored in the Laboratory in which they are generated, not in centralized accumulation areas. Right at the Laboratory designing phase especially for bigger Laboratories, sufficient space or facilities such as storage cabinets with partitions should be provided so that incompatible waste and non-waste chemicals or gases can be physically separated and stored. Laboratory waste water lines should be separate from domestic sewage, and if possible or necessary, a sampling point should be installed in an easily accessible location outside the building. This sampling point could be installed at a location where all the Laboratory wastes are discharged, before the Laboratory waste line connects to the domestic waste line. The sampling point should not be located in an area where water from irrigation or flow from storm water runoff can accumulate.

Within the Laboratory, wastes should be categorized and dumped in separate well labeled containers. Separation could be done simply by distinctive colours for each thrash can/basket with respect to waste category (**figure below**) e.g. blue for waste papers and green for plastics.

Laboratory waste segregation

Environmental impacts of Laboratory wastes and operations

The use of Laboratory equipment has many direct and indirect environmental impacts. Some direct impacts include:

- the considerable consumption of water, consumables and other resources;
- Whole Life Costing (WLC) which involves costs of consumables and their disposal and end of life disposal costs e.g. cost and annual kWh per litre of storage capacity for fridges and freezers,

- lifetime energy costs e.g. very high consumption of electricity [Source: S-Lab research];
- creation of waste, both in use and at end of life especially in the case where some equipment may require special and often expensive disposal due to the fact that they are contaminated.

There is also a large indirect impact from equipment-related requirements for floor space as building operation has considerable environmental impacts and, in some cases special requirements for building services such as constant temperature or humidity. The production of Laboratory equipment also has considerable environmental impacts although these are often hard to quantify. Minimizing these impacts by Laboratories is important for environmental health and safety.

Chapter 5

Procurement of Laboratory Supplies

Laboratory goods and services

The Laboratory management efficiency in terms of supplies lies in the existence of a procurement strategy whose aim should be to set out the planned approach for cost-effective purchasing of required goods and services, taking into consideration several factors such as the timeline for procurement, the funding and budget and the projected risks and opportunities. In most instances the strategy addresses the need to focus on the highest volume/value categories of goods and services purchased. It is important that Laboratories undertaking significant procurement have such an effective procurement strategy which clearly differentiates between procurement of support goods and services (such as support staff and office supplies), from those required to deliver the overall aims and objectives (including operational staff and payments to third parties for service delivery) of the Laboratory. The procurement strategy describes the contribution that effective procurement will make to the fulfillment of the organization's vision and objectives. It should also aim to identify the procurement approach to be followed for various categories of goods and services required, reflecting the comparative buying power that the organization possesses, with the strength of the market.

The Laboratory management could be more efficient if it has defined and documented policy and procedures for selection and use of purchased external goods and services in addition to an inventory control system for supplies. The Laboratory could establish appropriate quality records of external services, supplies and purchased products and should always evaluate suppliers of critical reagents, supplies and services that affect the quality of examination, and maintain records of these evaluations with a list of approved suppliers (Panadda, 2007). Value for money (VFM) is achieved by obtaining the right goods and services of the right quality, at the right time, to be delivered at the right place and at the right cost.

Procurement is the acquisition of goods and services at the best possible total cost of ownership at the appropriate quality, and at the right time. Procurement plays a key role in achieving VFM. The principles of transparency and the ability to demonstrate delivery of VFM, rather than simply referring to the existence of VFM policies are central to the procurement role. Buying goods and services more efficiently can release funds that could be used either to procure more goods and services or deliver other outputs. This entails that the Laboratory should have key VFM and best procurement practice principles. Stakeholders need to develop better understanding and better articulation of costs and results so as to make more informed, evidence-based choices in order to be sure they get the necessary quality at the lowest price.

Laboratory supply chain management

A supply chain (SC) includes all stages involved in fulfilling a customer request and is made up of the people, activities, information and resources involved in moving a product from supplier to customer. SC includes planning, purchasing, transportation, warehousing etc. Good SC management helps to reduce costs, increase customer satisfaction, and better utilize assets. Supply chain management (SCM) involves what is done along the SC in order to get the expected results. SCM is driven by the need to remove inefficiencies, excess costs and excess inventories from the supply pipeline. There are three levels of SCM; strategic, tactical and operational. The most important aspect of SCM is the planning, implementation and control of the SC.

Supply chain strategies require a total systems view of the linkages in the chain that work together efficiently to create customer satisfaction at the end point of delivery to the consumer.

Efficient Laboratory supply chain management is also determined by the Laboratory organization and management support realities. The Laboratory setting and organigramme should be such that needs are clearly identified and neatly presented by an officer in charge (Laboratory Manager, Principal Investigator, etc.) to the financial department or management of the institution through the Laboratory Financial Officer. Once approved, supply or procurement should be done with the involvement of the end users (Laboratory staff) in order to ensure that the right product of the right quantity and in the right condition is purchased with the right documentation from and for the right place at the right time and right price.

There is also the need for good collaboration with key suppliers to avoid any changes in supply and schedules. Management support is vital in this instance. Management could improve SC performance by shifting more risk to suppliers. This entails good demand and supply planning to keep SC as lean as possible even though SCs need to be treated as one integrated organization with support from all SC functions. Here, information is important in integration including demand signals, forecasts, inventory, transportation, etc. Biosafety and Biosecurity issues must equally be integrated in the SC to ensure safe handling, transportation and storage of goods. Furthermore, all SC activities must comply with relevant environmental regulations (hazardous material, gases, liquids etc) to reduce environmental pollution and enhance sustainable sourcing.

Chapter 6

Laboratory equipment operation

Laboratory safety and efficient management are enhanced by the way the Laboratory equipment are operated. Laboratory operators need to have an overview of different categories and types of Laboratory equipment at their disposal with knowledge of their safe handling, operation and maintenance following well set schedules.

Categories of Laboratory equipment

The various categories of equipment commonly used in Laboratories can be classified with respect to their usage. The following is a non exhaustive list outlining some of the general categories of Laboratory equipment:

i. *Analytical equipment* which are used to analyze samples, or their components, and record or output data specific to the initially set parameters. They include chromatographs (Gas Chromatograph, High Performance Liquid Chromatographs, etc), spectrophotometers, DNA Sequencers, etc.

ii. *Heating or thermal processing equipment* used for raising sample temperatures or controlling and maintaining environmental temperatures commensurate with preset reference levels such as sterilizers like the autoclaves, incubators, burners, water baths, ovens, growth chambers, etc.

iii. *Measuring and test equipment* used for conformity measurement of parameters like weight, volume, wavelength, radiation intensities, etc or to test for levels of specific components in a sample e.g. electronic balances, pH meters, pipettes, volumetric flasks, microscopes, etc.

iv. *Mixing and separation equipment* are those used for mixing and separation of samples. They include instruments such as the centrifuges, magnetic stirrers, shakers, etc.

v. *Standby equipment* are the category used to address emergency situations and to ensure continuity of activities in the Laboratory environment. These include standby material like freezers, emergency circuits, standby generators and the like.

vi. *Equipment for material storage* used for the storage of samples and consumables, especially those that are temperature sensitive. They include cupboards, freezers, cryogenic tanks, cold rooms and gas storage tanks, etc.

vii. *Safety equipment* which must be present for the safety of Laboratory users and materials. Some may serve to prevent or limit sample contaminations. Others may serve to mitigate Biosafety risks. Examples of this category include fume hoods, alarm systems, safety cabinets, etc. Laboratory access control like special doors and the like fall under this category.

viii. *Utilities* are the category of equipment that mainly support research functions such as water purifiers, water deionizers, liquid nitrogen plant, waste disposal equipment and plants etc.

ix. *Environmental and comfort Equipment* are the category of Laboratory equipment used to enhance Laboratory ambiance as well as maintain a conducive environment for staff and life specimens when present. Most environmental and comfort equipment are those used for monitoring and control of changes in the Laboratory environment. They include humidifiers, light level sensors and equipment, smoke detectors, HVACs, exhaust fans, etc.

All the above categories of equipment can function well and better serve their purpose if and only if they are safely handled. Safe handling enhances good equipment operation, a longer life span, reliability, and safety to the operator. To this effect, user training is a very important aspect that must be included within the sales and/ or purchase agreement relative to each major Laboratory equipment. Operators should learn and effectively apply good Laboratory practices relative to equipment operations. The user and operation manual of Laboratory equipment are to be well exploited for the same purpose. Labels and safety features on the equipment must be taken into consideration when in use. A safety or engineering department should be in place to ensure that best practices are maintained in handling of all equipment. The equipment or machines should be operated as per the manufacturer's instructions and recommendations. Management should establish and follow-up an annual preventive maintenance schedule for Laboratory equipment. The efficiency of most equipment relies in proper calibration and efficient setting of necessary parameters on the equipment before running them as well as implementation of the standard operation procedures (SOPs) for the particular machine. Moreover, the consistency of data generated by most analytical equipment is determined by the relevant method development and validation processes that precede analysis proper.

Laboratory equipment maintenance

Proper maintenance of Laboratory equipment is a constitutive element of Good Laboratory Practice. It is good practice for any research or clinical Laboratory to have an internal or external equipment maintenance service provider. Laboratory equipment maintenance is necessary for assays to function within manufacturer's specifications (Ezzelle *et al.,* 2008). Laboratory staff must conduct maintenance and service per manufacturer specifications by following documented daily, weekly, and/or monthly routine maintenance plans for all equipment utilized to ensure that all equipment performs consistently and reproducibly during the conduct of the trial (Ezzelle *et al.,* 2008). The laboratory must equally document all scheduled maintenance, unscheduled maintenance, service records, and calibrations for all equipment (Ezzelle *et al.,* 2008).

Laboratory equipment maintenance can be manual, automated with mounted sensors for example, or computer based such as remote monitoring through internet. The Laboratory maintenance unit must have staff that are well trained, and who can work with a good support from the management. There are several categories of equipment maintenance some of which are as follows:

i. *Corrective or reactive maintenance* in which case the Laboratory initiates a request for a given maintenance service after a machine breakdown.

ii. *Preventive or pro-active maintenance* which involves the service provider planning in advance and requesting the client to render available the machine for service. Internal preventative maintenance activities as well as vendor provided maintenance/repair for laboratory equipment is paramount in providing accurate and reliable results (Ezzelle *et al.,* 2008).

iii. *Predictive maintenance* involves the measurement and detection of the onset of equipment or system breaking down such as equipment misbehavior; like fans and bearing noises, erratic temperatures, etc.

iv. *Inspection and calibration* maintenance usually involves the invitation of specialized firms or Government recommended inspectors or service providers, to service or inspect, calibrate, certify, and validate the performance of Laboratory equipment in order to conform to the manufacturer's set standards of operation or to comply with a legal requirement.

v. *Operational maintenance* is carried out by the user as stipulated in the user manual e.g. cleaning, testing or calibration before use to ensure that the machine performs optimally and guarantees reproducibility of results.

A Laboratory may engage an equipment service contract with suppliers. In such a case, pre-qualification of suppliers is important so as to identify service providers in advance. Several reasons may justify the need to engage external service providers for the maintenance of Laboratory equipment. The Laboratory might have specialized equipment with complex configuration, new technology, or without internal capacity. External service providers may also be solicited due to legal requirements such as government standards, legal and safety requirements, or equipment warranty after purchase. More still, this could be due to specialized services that require empowerment of the internal engineering team so as to reduce such costly services in the future. Another reason could be due to legal compliance or requirements. Whatever the reason, the service contract is best negotiated during the procurement of the equipment.

Chapter 7

Laboratory information management system

The Laboratory information management system (LIMS) or Laboratory Information System (LIS) is an important and integral part of Laboratory operations relevant for efficient Laboratory management. The LIMS or LIS is a powerful tool to manage complex processes, ensure regulatory compliance and promote collaborations between multiple laboratories. Usually a LIS is capable of consolidating disparate scientific processes into a single, compliant platform with comprehensive reporting, surveillance and networking capabilities. The result of proper LIS usage is vastly enhanced data management and data sharing-within the Laboratory and across Laboratories (Ezzelle *et al.,* 2008).

The use of LIMS must be compliant with regulatory standards that affect the Laboratory such as ISO/IEC 17025, ISO 15189, and Good Laboratory practice guidelines. The base set of functionality that defines LIMS can roughly be divided into five Laboratory processing phases, with numerous software functions falling under each (Anonymous, 2012). They include:

- The reception and log in of a sample and its associated customer data;
- The assignment, scheduling, and tracking of the sample and the associated analytical workload;
- The processing and quality control associated with the sample and the utilized equipment and inventory;

- The storage of data associated with the sample analysis;
- The inspection, approval, and compilation of the sample data for reporting and/or further analysis.

Core functionalities associated with LIMS

In a Laboratory information management system, several core functionalities are associated with each Laboratory processing phase. Some of the general core functionalities are: electronic data exchange, sample management, instrument and application integration etc.

Electronic data exchange: The exponentially growing volume of data created in Laboratories coupled with increased business demands and focus on profitability have pushed LIMS vendors to increase attention to how their LIMS handle electronic data exchanges. Vendors now pay much attention to the way the input and output data is managed. A very important aspect of the modern LIMS is the successful transfer of data files in Microsoft Excel and other formats, as well as the import and export of data to Oracle, SQL, and Microsoft Access databases (Anonymous, 2012).

Sample management: The core function of LIMS is traditionally the management of samples (Anonymous, 2012). The usual situation is that the Laboratory worker matches samples to documents. With a LIMS, sample management in the Laboratory is made more efficient since the manual matching of samples to documents is replaced by the LIMS automated system in which each sample received in the Laboratory is registered. It is possible for a customer to place an order for a sample directly to a LIMS. The processing could then include barcode generation for the sample and registration of the sample container. Various parameters such as the barcode to be affixed on the sample container, clinical or phenotypic information corresponding with the sample are also often recorded. The LIMS then tracks chain of custody as well as sample location. Location tracking may involve assigning a sample to a particular freezer location, often down to the granular level of shelf, rack, box, row, and column.

Instrument and application integration: It is possible to integrate modern LIMS with Laboratory instruments and applications. A LIMS may create control files that are fed into the instrument to direct its operation on some physical item such as a sample tube or sample plate. The LIMS may then import instrument result files to extract data for quality control assessment of the operation on the sample. Access to the instrument data can sometimes be regulated based on chain of custody assignments or other security features if need be (Anonymous, 2012).

Audit management: This functionality of LIMS helps to fully track and maintain an audit trail.

Barcode handling through which a Laboratory operator can assign one or more data points to a barcode format and/or read and extract information from a barcode.

Chain of custody is the functionality which also helps to assign roles and groups that dictate access to specific data records and who is managing them.

Compliance functionality helps to follow regulatory standards that affect the Laboratory.

Customer relationship management is the LIMS operation which helps to handle the demographic information and communications for associated clients.

Document management is the LIMS functionality which helps to process and convert data to certain formats. It can also help to manage how documents are distributed and accessed.

Instrument calibration and maintenance is the LIMS functionality which helps to schedule important maintenance and calibration of Laboratory instruments and to keep detailed records of such activities.

Inventory and equipment management measures and records inventories of vital Laboratory supplies and equipment.

Manual and electronic data entry is that functionality which provides fast and reliable interfaces for data to be entered by a human or electronic component.

Method management functionality provides one location for all Laboratory process and procedure (P&P) and methodology to be housed and managed as well as connecting each sample handling step with current instructions for performing the operation.

Personnel and workload management is a LIMS functionality which helps to organize work schedules, workload assignments, employee demographic information, training, and financial information.

Quality assurance and control is a LIMS functionality which gauges and controls sample quality, data entry standards, and workflow.

Reports functionality of LIMS helps to create and schedule reports in a specific format and to schedule and distribute reports to designated parties.

Time tracking calculates and maintains processing and handling times on chemical reactions, workflows, and more.

Workflows: This is the LIMS functionality which helps in tracking a sample, a batch of samples or a lot of batches through its lifecycle.

Traceability is the LIMS functionality which helps to effectively manage and trace samples.

LIMS architectures

LIMS architecture actually represents the way the LIMS is configured and the way it will function to handle a particular issue. Some LIMS architectures include : thick-client LIMS; thin-client LIMS; hybrid LIMS architecture that incorporates the features of thin-client browser usage with a thick client; web-enabled LIMS architecture which is a thick-client architecture with an added web browser component; and web-based LIMS architecture which is a hybrid of the thick- and thin-client architectures considered as one of the most confusing architectures.

Thick-client LIMS

The thick-client LIMS was one of the first architectures implemented into a LIMS. It is a client/server architecture with some of the system residing on the computer or workstation of the user or client and the rest on the server which has the primary purpose of data storage. The LIMS software is installed on the client computer, which does all of the data processing. Later it passes information to the server for storage. Most changes, upgrades, and other modifications are done on the client side. The thick-client LIMS has the advantage of providing higher processing speeds since processing is done on the client side and not on the server. It is slightly more secured given that access to the server data is limited only to those with client software. Thick-client systems have the added advantage of providing more interactivity and customization. The thick-client LIMS can become web-enabled through an add-on component (Anonymous, 2012).

Thin-client LIMS

Thin-client LIMS is an architecture which offers full application functionality accessed through a device's web browser. With thin-client LIMS, the actual LIMS software resides on a server or host which feeds and processes information without saving it to the user's hard disk. Any necessary changes, upgrades, and other modifications are handled by the entity hosting the server-side LIMS software. A thin-client LIMS leaves no footprint on the client's computer and only the integrity of the web browser needs to be maintained by the user. The advantages of this system include significantly lower cost of ownership and fewer network and client-side maintenance expenses. Another implementation of the thin client architecture is the Maintenance, Support, and Warranty (MSW) agreement for which pricing levels are typically based on a percentage of the license fee, with a standard level of service for 10 concurrent users being approximately 10 hours of support and additional customer service (Anonymous, 2012).

Hybrid LIMS architecture

There exists a hybrid architecture of LIMS in the form of a web-based LIMS that incorporates the features of thin-client browser usage with a thick client. Some LIMS vendors rent hosted, thin-client solutions as "software as a service" (SaaS). These solutions are less configurable than on premise solutions and are therefore considered for less demanding implementations such as Laboratories with few users and limited sample processing volumes. (Anonymous, 2012).

Web-enabled LIMS architecture

A web-enabled LIMS architecture is a thick-client architecture with an added web browser component. In this setup, the client-side software has additional functionality that allows users to interface with the software through their device's browser. The advantage of a web-enabled LIMS is that end-users can access data both on the client side and the server side of the configuration. As in a thick-client architecture, updates in the software must be propagated to every client machine (Anonymous, 2012).

Web-based LIMS architecture

Web-based LIMS architecture is a hybrid of the thick- and thin-client architectures. With a web-based LIMS, much of the client-side work is done through a web browser. Web-based architecture has the advantage of providing more functionality through a friendlier web interface (Anonymous, 2012). However, this LIMS architecture is considered as one of the most confusing architectures.

Since each Laboratory's needs for tracking additional data points can vary widely modern LIMS are conceived with extensive configurability; vendors create LIMS that are adaptable to individual environments. Whatever the architecture, reports generated by the LIMS, and those created by other means, must be consistent in format and readability and should imperatively include the following information (Ezzelle *et al.*, 2008):

- The origin of the specimen (geographical location), transportation means and conditions;
- the study participants' names and/or a unique identifier;
- the name and address of the laboratory location where the assay was performed;
- the date and time of specimen receipt into the laboratory;
- the assay report date;
- the name of the test performed;
- specimen source; (e.g. from fruits, blood etc) ;
- the assay result and if applicable, the units of measurement or interpretation or both;
- reference ranges (normal, Analytical Measurement Range (AMR) or the Clinically Reportable Range (CRR);
- any information regarding the condition and disposition of specimens that do not meet the laboratory's criteria for acceptability;
- the records and dates of all assays performed.

Written SOPs for the operation of the LIMS containing the purpose, the way it functions, and its interaction with other devices or programs as well as a disaster-preparedness plan (for the preservation of data and equipment) must be available in the Laboratory and should be appropriate and specific to the current activities of all laboratory staff. The documentation must be consistent with validation data and results including data entry, data transmission, calculations, storage and retrieval. All users of the LIMS system should receive adequate training both initially and after system installation (Ezzelle *et al.*, 2008).

Some limitations of LIMS

The use of LIMS is not without shortcomings. The deployment of a LIMS solution outside the firewall of an organization renders the LIMS liable to potential intrusion. Web-based and web-enabled deployment of LIMS is liable to possible exploitation by hackers where sensitive Laboratory and research data may be compromised. Moreover, LIMS implementations might sometimes be lengthy and costly (Royce and John, 2010). This is due in part to the diversity of requirements within each Laboratory and also to the inflexible nature of LIMS products for adapting to these widely varying requirements. Thick-client systems often require a greater learning curve for users. Moreover, client-side LIMS require more robust client computers and more time-consuming upgrades, as well as a lack of base functionality through a web browser. On its part, thin-client LIMS architecture has the disadvantage of requiring real-time server access, a need for increased network throughput, and slightly less functionality.

For web-enabled LIMS architecture, the disadvantages of requiring always-on access to the host server and the need for cross-platform functionality entail additional costs to end users. Web-based architecture has the disadvantages of being more costly in system administration and support for Internet Explorer and .NET technologies, and reduced functionality on other devices.

Due to these shortcomings, newer LIMS solutions now take advantage of modern techniques in software design and can be configured and adapted more easily than prior solutions. Furthermore, their implementations are much faster at lower costs and with minimized risk of obsolescence.

A case study of LIMS

As mentioned earlier, good Laboratory information management requires appropriate software. Some examples of vendors that develop and market LIMS include Thermo Fisher Scientific, AgileBio (LabCollector), Analytik Jena, LabLynx, Labvantage, PerkinElmer, Sapio Sciences, STARLIMS, Waters Corporation, and Ocimum Bio Solutions.

LabCollector is one of such consistent LIMS for modern Laboratories. It is a network based system like an Intranet. LabCollector, a product from AgileBio is a totally web-based application. It can permit access and management of a great variety of information in the Laboratory. LabCollector can be installed on one of the Laboratory computers which will play the role of a server for the remaining computers in the Laboratory. The use of web technology makes it a light solution, as no client application has to be installed on each computer. It is cross-platform and so can be installed on any operating system (Windows, Mac OSX, Linux …). LabCollector can be accessed from all computers from the same local network. It can even be accessed through the Internet. Therefore, data access and management can be password protected. The interface is accessed through simple Internet browsers like Internet Explorer, Firefox, Safari etc. Therefore, with LabCollector, Laboratory information and data is reachable from anywhere and close to where it is needed. The support of wireless devices brings even more mobility. LabCollector can be gotten simply by visiting the website at www.labcollector.com for details.

Chapter 8

Quality Control and Quality Assurance

Quality control and quality assurance constitute good laboratory management practices. There are several definitions for quality. In general, Quality entails providing appropriate services for appropriate people at the appropriate time and with efficient practical procedures according to average strength of society (Rabinson, 1999; Dargahi and Rezaiian, 2007). The International Standardization Organization (ISO) defines quality as a set of characteristics of a service or product which provide the customer's requirements (Dargahi and Rezaiian, 2007). Quality is therefore the measure of excellence or state of being free from defects, deficiencies or significant variation. This is feasible through strict and consistent adherence to measureable and verifiable standards to achieve uniformity of products that satisfy specific customer needs. Quality Control (QC), Quality Assurance (QA) and Total Quality Management (TQM) are the three major elements of quality (Dargahi and Rezaiian, 2007).

Quality Control

Quality control (QC) is a process employed to ensure a certain level of quality in a product or service. The basic goal of quality control is to ensure that the products, services, and processes provided meet specific requirements and are dependable and satisfactory. QC involves the examination of products, services or processes for certain minimum levels of quality as well as identification of products that do not meet certain specified standards. There are specific technical activities used to appreciate the quality of a product.

The efficiency of Laboratory operations also lies in the existence of a QC program put in place by the Laboratory Manager. The quality control program supports functions in the areas of Test standards and controls, reagents, test specimens, review of quality control data, quality control logs, labeling of quality control materials and reagents, inventory control, parallel testing, and water quality testing. Such a QC program helps to detect immediate errors and changes that occur over time thereby assuring the accuracy and reliability of test results (Ezzelle *et al.,* 2008). The Manager is also expected to determine the number and frequency of QC tests, as well as the appropriate QC materials to use (Ezzelle *et al.,* 2008).

Internal quality control (IQC) procedures have an important part to play in monitoring the day to day performance of assays used in the Laboratory (Gray *et al.,* 1995). Laboratories could include internal quality control (IQC) measures with acceptable limits in order to increase confidence in the validity of their results (Gray *et al.,* 1995). Quality control measures permit the assessment of accuracy and precision in results obtained in the laboratory. In clinical Laboratories for instance, inclusion of such measures is designed to increase the probability that assay results reported by the Laboratory are valid and that clinicians may confidently use those results when making diagnostic or therapeutic decisions (Gray *et al.,* 1995). Internal quality control samples should therefore be included in the assays performed in the Laboratory. When the results lie within pre-determined limits, they could then be used to validate the test results (Gray *et al.,* 1995).

Quality system

The ISO defines a quality system as the organizational structure, responsibilities, procedures and resources needed for implementing quality management. The quality system aims at ensuring the consistency, reproducibility, traceability and reliability of the products or services (Panadda, 2007). A quality system has the following five key elements: Organizational management and structure; Quality standards; Documentation; Training; and Assessment (Panadda, 2007). These key elements can be referred to in the various existing International Standardization Organization (ISO) or certification/ accreditation schemes. For instance, the ISO 9000 series presents a collection of good management practices related to quality systems and is composed of generic and specific standards. The ISO 9001- 2000 "Quality Management System Requirement" presents requirements to the implementation of Quality Management and Quality Assurance (Perigo and Rabelo, 2005; Dargahi and Rezaiian, 2007). On its part, ISO 15189: 2003, Medical Laboratories - Particular requirement for quality and competence (Aoyagi, 2004), provides a framework for the design and improvement of process - based Quality Management by Medical Laboratories (Dargahi and Rezaiian, 2007; Kubono, 2004). Moreover, ISO 15189 requirements consist of two parts, one is management requirement and the other is technical requirement. The latter includes the requirements for Laboratory competence such as personal, facility, instrument and examination methods (Aoyagi, 2004; Dargahi and Rezaiian, 2007; Dargahi and Rezaiian, 2007).

Quality assurance

Quality assurance (QA) is defined as a set of quality principles that represent the foundation for continuous organizational improvement. It is a management system that has customer satisfaction as a prime objective (Dargahi and Rezaiian, 2007). QA aims at monitoring and evaluating a system or Laboratory to ensure that standards and quality are met. QA is a long term process which requires leaders to maintain their commitment, keep the process visible, provide necessary support and maximize employees' involvement in design of the system (Dargahi and Rezaiian, 2007). QA should therefore be a planned and systematic process for monitoring and evaluating the quality and appropriateness of Laboratory services. It should be supplemented by Quality Improvement (QI) and other quality methods (Dargahi and Rezaiian, 2007).

Efficient QA implementation depends on recognition of the importance of interdisciplinary and/or multidisciplinary collaboration of the Laboratory staff (Orgam, 1995; Dargahi and Rezaiian, 2007). In an efficient QA system, employees recognize their responsibilities and perform their duties with group participation and decision making. In such a system, all employees are consciously committed to QA implementation (Dargahi and Rezaiian, 2007). Implementation of QA in a Laboratory could be through application of ISO standards, conception and utilization of a quality manual and eventual accreditation. The Laboratory Quality Manual helps to efficiently communicate to the laboratory staff the manner in which laboratory testing is to be conducted. Adherence to the Quality Manual by laboratory staff is essential to ensure both the quality and consistency of results generated (FAO, 2013).

Advantages of a QA system include increased customer confidence, improved institutional credibility; improved work processes and efficiency; provision for a planned system of review of procedures usually conducted by third parties, etc. Laboratory accreditation is an efficient tool for putting in place a quality system toward achieving continuous improvement in Laboratory services in a sustainable way (Panadda, 2007). In fact, accreditation is a tool that recognizes the existence of a quality system in a Laboratory (Panadda, 2007).

Total Quality Management

The development of Total Quality Management (TQM) from 1950 onwards can be credited to the works of various experts among whom, Edward Deming, Joseph Juran and Philip Crosby who have contributed significantly towards the continuous development of the subject (Marya, 2006 ; Alejandro, 2011). In fact, Dr. W. Edwards Deming, regarded as the "father" of post war Japanese economic miracle as well as leading advocate of the TQM movement in the

United States, developed a systematic approach to solving quality related problems which aims to fulfill customer expectations (Phu Van Ho, 2011).

According to Brian *et al.,* (1989) TQM is a means for improving personal effectiveness and performance and for aligning and focusing all individual efforts throughout an organization. TQM is a continuous quality improvement process that evaluates processes from a customer satisfaction point-of-view. This approach is focused on satisfying customers' expectations, identifying problems, building commitment, and promoting open decision-making among workers (Arnold *et al.,* 1992). It is a participative, systematic approach to planning and implementing a continuous organizational improvement to achieve and exceed standards quickly and efficiently (Arnold *et al.,* 1992). Therefore, the aim of TQM is continuous process improvement (Simpson *et al.,* 1991).

The product quality and manufacturing process of suppliers has great effect on the quality of the final product (Guangshu, 2009). TQM approach is based on a set of management practices and statistical measures that, when combined, can remove the causes of poor product quality and excessive cost (Houston and Dockstader, 1988). This approach emphasizes the major role that managers have in achieving quality and productivity improvement for an organization. It is in this vein that Deming and other TQM proponents, such as Crosby and Juran, estimate that up to 85 percent of quality improvement is under direct control of management (Houston and Dockstader, 1988).

The series standards of ISO9000 relative to TQM are implemented in many industries, such as manufacturing, service, health care, nonprofit organizations, educational institutions, and even public bureaucracies (Guangshu, 2009; Noori, 2004). With time, the emphasis of research and practice of TQM has transferred from enterprise focus to supply chain focus, involving not only the high quality of products and services but also the high level of quality control of the whole system. The implementation of total quality management (TQM) in a supply chain system is determinant for the survival of enterprises. There are certain modern TQM principles of ISO9000 in supply chain quality management, namely customer focus, leadership, involvement of people, process management, system management, continual improvement, factual approach to decision-making, and mutually beneficial supplier relationships (Guangshu, 2009). The establishment of TQM system in a setting based on the management ideas of ISO9000 can efficiently promote the involvement of all the members and facilitate the implementation of quality control of the whole system (Guangshu, 2009).

Health-care organizations are now applying the principles of TQM (Westgard *et al.,* 1991). The integration of TQM into public health agencies complements and enhances the Model Standards Program and assessment methodologies, such as the Assessment Protocol for Excellence in Public Health (APEX-PH), which are mechanisms for establishing strategic directions for public health (Arnold *et al.,* 1992).

TQM applies analytical or group problem-solving tools and techniques, such as process flow charts, check sheets, run charts, brainstorming, nominal groups, fishbone diagrams, consensus, and Pareto[5] analysis in new ways to facilitate communication and decision making (Simpson *et al.,* 1991; Westgard *et al.,* 1991; Arnold *et al.,* 1992). Laboratories are particularly suited to TQM because laboratorians are familiar with the TQM tools even though they must learn to apply these tools to new areas (Simpson *et al.,* 1991). Implementing TQM in a laboratory requires incorporating quality improvement (QI) and quality planning (QP) with Quality Laboratory Practices (QLP), Quality Control (QC) and Quality Assurance (QA) to provide a complete quality management system (Westgard *et al.,* 1991). Because of its great importance and potential efficiency, we strongly recommend the worldwide introduction of TQM as a subject in the graduate and postgraduate curriculum of universities (Marya, 2006) as is already the case in several countries.

---

5 Pareto analysis is a formal technique useful where many possible courses of action are competing for attention. In essence, the problem-solver estimates the benefit delivered by each action, then selects a number of the most effective actions that deliver a total benefit reasonably close to the maximal possible one (Source: http://en.wikipedia.org/wiki/Pareto_analysis accessed 10/04/2015).

Chapter 9

Laboratory Accreditation

The increased demand for testing has to be matched with high quality testing; accreditation is one means of assuring continuous quality testing services (Herman *et al.,* 2004). In the global trade system, the buyer would like to purchase goods with the specified quality if it has been tested by an accreditated Laboratory (Panadda, 2007). Laboratory accreditation could also be considered a means to show case legitimacy and credibility. According to Abrahamson (1991) cit. Jeffrey and Stefanie (2009), organizations embrace management trends adopted by peer organizations in order to appear legitimate by conforming to norms. Accreditation of a Laboratory is the process by which an independent and authorized agency accredits the quality system and competence of a Laboratory on the basis of certain pre-defined standards (Panadda, 2007). It is the formal recognition, authorization and registration of a Laboratory that has demonstrated its capability, competence and credibility to carry out the tasks it is claiming to be able to do. Accreditation provides feedback to Laboratories as to whether they are performing their work in accordance with international criteria for technical competence (Kanagasabapathy and Pragna, 2005). Laboratory accreditation is also a means to improve customer confidence in the test reports issued by the Laboratory (Kanagasabapathy and Pragna, 2005).

Accreditation process essentials and requirements

The Laboratory accreditation system is important for the acceptance of test results nationally and internationally. Panadda (2007) provided guidelines on the facilities and personnel needed, and examples of how to initiate the establishment of an accreditation process in a system with the goal of making a beginning with national standards and achieving an internationally acceptable accreditation system. To achieve recognition, a Laboratory must first implement the requirements of ISO (Panadda, 2007). Accreditation is done at regular intervals to ensure maintenance of standards and reliability of results generated by the Laboratory (Panadda, 2007). Medical Laboratories use a different international standard, ISO 15189, which is specially formulated for medical Laboratories only (Aoyagi, 2004), while all other types of Laboratories including calibration and testing Laboratories use ISO/IEC 17025 (Panadda, 2007).

The three essential guiding principles for establishing accreditation are: setting up an accreditation body; formulation or adoption of standards; and implementation mechanism (Panadda, 2007). Kanagasabapathy and Pragna (2005) acknowledge the importance of Laboratories making a definite plan for obtaining accreditation and nominating a responsible person as Quality Manager, who should be familiar with the Laboratory's existing quality system, to co-ordinate all activities related to seeking accreditation. The same authors have outlined the procedures for accreditation of clinical Laboratories by The National Accreditation Board for Testing and Calibration Laboratories (NABL), an autonomous body under the aegis of the Department of Science and Technology, Government of India, according to ISO 15189 (Kanagasabapathy and Pragna, 2005).

It is important to choose the best suited accreditation and/or certification system depending on the purpose of the Laboratory in place (Dargahi and Rezaiian, 2007; Tazawa, 2004). Accreditation is however an expensive exercise and many non-commercial organisations usually opt not to proceed with accreditation but put in place good QC and QA principles (Kanagasabapathy and Pragna, 2005).

Implementation of accreditation at national level

The coordination of partner support and good leadership and ownership of the state are prerequisites for successful and coordinated Laboratory management, classification and accreditation. The state has the responsibility of providing services to citizens such as comprehensive diagnostic testing of all prevalent major infections, monitoring of patient treatment, drug resistance testing and surveillance studies that inform policymaking decisions and major health reforms (Mothabeng *et al.,* 2012). Countries can adapt existing international standards for Laboratory accreditation or formulate theirs (Panadda, 2007). Whatever the case, the accreditation process requires identification of an authoritative body, adoption of standards, and institution of a mechanism of assessment of the Laboratory to certify its compliance with standards (Panadda, 2007). In the accreditation process, the ISO17011 standard published by the ISO and entitled *General requirements for accreditation bodies accrediting conformity assessment bodies* is usually followed by accreditation bodies to get recognition from regional and international bodies (Panadda, 2007). The standard is composed of clauses which include the scope, normative references, terms and conditions, accreditation body, management, human resources, accreditation process, and responsibility of the accreditation body to the conformity assessment body. Implementation of accreditation at the national level requires a step-wise approach and national commitment (Panadda, 2007) starting with the development of national policy, establishment of accreditation body, formulation and dissemination of national standards, development of process of accreditation and its dissemination, and forging linkages with national and international partners.

For efficient implementation of this step-wise approach, it is important for each country to define national Laboratory policy operational plans that directly address issues of Laboratory management, classification and accreditation. An interesting example is the Lesotho Ministry of Health and Social Welfare (MOHSW) which is committed through a strategic plan to fulfilling such mandate (Mothabeng *et al.,* 2012). The MOHSW adopted the World Health Organization Regional

Headquarters for Africa's Stepwise Laboratory Quality Improvement Toward Accreditation (WHO–AFRO–SLIPTA) process to replace the Strengthening Laboratory Management Towards Accreditation (SLMTA) programme across the whole country, becoming the first African country to do so. Performance of this system could be tracked using the WHO–AFRO-SLIPTA checklist, with assessments carried out at baseline and at the end of SLMTA. In the Lesotho experience, two methods were used to implement SLMTA: the traditional 'three workshops' approach and twinning SLMTA with mentorship that involves intensive follow-up visits. In such a process, areas of improvement towards accreditation may include corrective actions, documents and records, and process improvement (Mothabeng *et al.*, 2012).

National accreditation standards, a step towards conformity with international standards

Development of a national standard as a starting standard for any country, especially developing countries, is one of the logical ways of initiating and implementing an accreditation programme. The standard may differ from one country to another depending on the state of development of the quality system in Laboratories. The national standard must be aligned with the international standard, e.g. in the case of medical Laboratories, the aim of accreditation to ISO 15189 shall be the final target (Aoyagi, 2004; Panadda, 2007). National accreditation programmes should be established with the primary goal of strengthening the quality system in Laboratories and promoting quality improvement on a continual basis. In this light, the following issues could be taken into consideration in the process of establishing such a programme (Panadda, 2007):

- Identification of standards that will be used in the country;
- Establishment of an accreditation body with adequate resources, infrastructure and personnel;
- Characterization of the concepts of the accreditation body;
- Definition of areas that will be covered by the scheme;

- Identification of international accreditation bodies (e.g. **Table 1**) to accompany the process;
- Identification of stakeholders: the government, private sector, regulatory agency, Laboratory professionals and users;
- Identification of customers and delineation of the relationship with them;
- Setting up a process of accreditation and publicizing it widely.

Table 1. International accreditation agencies*

Agency	Membership	Organization and role
International Laboratory Accreditation Cooperation (ILAC); apex international authority on Laboratory accreditation	Accreditation bodies and affiliated organizations throughout the world.	The accreditation bodies in regional arrangements are responsible for maintaining the necessary confidence in their respective areas.
Asia Pacific Laboratory Accreditation Cooperation (APLAC)	The APLAC groups accreditation bodies in the Asia Pacific region.	Fosters the development of competent Laboratories and inspection bodies in participant countries, harmonizes accreditation practices within the region and with other regions, and facilitates mutual recognition of accredited (MRA) results through the APLAC multilateral MRA.

*Sources: Panadda Silva, 2007. Guidelines on Establishment of Accreditation of Health Laboratories. WHO Regional Office for South-East Asia, New Delhi, 49 p. **WHO publication Quality standards in health Laboratories, implementation in Thailand: a novel approach, World Health Organization Regional Office for South-East Asia, 2005 (SEA-HLM-386).

It is important for all Laboratories in each country to be well structured for easy coordination of the above issues (Mothabeng *et al.*, 2012). A case study is the Laboratory system in Lesotho which is structured in three tiers: referral, regional and district Laboratories. The

Central Laboratory at the Queen Elizabeth II (QE II) Hospital in Maseru serves as the national reference Laboratory. There are two regional Laboratories and 16 other Laboratories are at the district level, with 7 of these managed by the MOHSW, one by the military and 7 by the Christian Health Association of Lesotho. The final district Laboratory is owned by the Partners in Health, a non-governmental organization.

With inspiration from the Lesotho experience, key pillars that can support successful accreditation of Laboratories include (Mothabeng *et al.*, 2012):

- A national Laboratory policy;
- A time framed national strategic plan;
- The appointment of a national Laboratory Director;
- The establishment of a national Quality Assurance Unit (QAU) headed by a National Quality Manager who is supported by key National Quality Officers.

Countries that do not have a sound quality system in place may not benefit from accreditation until quality assurance in terms of good Laboratory practices, internal quality control (IQC), audit, validation, internal quality assessment and participation in external quality assessment schemes (EQAS) are strengthened (Panadda, 2007).

International standards for Laboratory accreditation

International standards used for Laboratory accreditation (**Table 2**) have evolved with time to suite the constantly changing exigencies of quality and credibility of Laboratory performance. In 1989, an international standard for Laboratories, ISO/IEC Guide 25 [*General requirements for the competence of calibration and testing Laboratories*] was published and used as a guidance document for many organizations throughout the world (Panadda, 2007). In 2005, the ISO/IEC Guide 25 standard was reviewed, to use the same wording as ISO 9001:2000 which is the standard for quality systems in general production and services (Panadda, 2007). The first edition of ISO/IEC

17025:1999 replaced ISO/IEC Guide 25 and EN 45001. Till 2006, two International standards, ISO/IEC 17025:2005 and ISO 15189:2003, were used as tools for Laboratory accreditation by accreditation bodies.

Table 2. Some international standards for Laboratory accreditation*

Standard	**Destined organization**	**Year of publication**
ISO/IEC Guide 25: 1990	Calibration and testing Laboratories	Published in 1989 and reviewed in 2005
ISO/IEC 17025	Calibration and testing Laboratories	First edition 1999; second edition 2005 and last revision in 2010
ISO 15189	Medical Laboratories only	Published in 2003, new version in 2012
ISO 9001	Could be implemented for issues about management system and not technical competence of the Laboratory	1994, 2000 and 2001
ISO 9002		1994

*Sources: Panadda Silva, 2007. Guidelines on Establishment of Accreditation of Health Laboratories. WHO Regional Office for South-East Asia, New Delhi, 49 p.

The ISO/IEC 17025 [*General requirements for the competence of testing and calibration Laboratories*] was intended to overcome the weak points of ISO/IEC Guide 25. This new standard has more details in each clause as compared with the old one and also has more content to suit Laboratory practices. It contains all the requirements that testing and calibration Laboratories have to meet if they wish to demonstrate that they operate a management system, are technically competent, and are able to generate technically valid results.

ISO 9001:1994 and ISO 9002:1994 are only related to issues about a management system in the first edition. They do not cover the issues related to technical competence. These standards have been superseded by ISO 9001:2000 in the second edition of ISO/IEC 17025:2005. ISO/

IEC 17025:2005, clause 1.6 states that "If testing and calibration Laboratories comply with the requirements of this international standard, they will operate a quality management system for their testing and calibration activities that also meets the principles of ISO 9001." But ISO/IEC 17025:2005 also stipulates that demonstrated conformity to ISO/IEC 17025 does not imply conformity of the quality management system within which the Laboratory operates to all the requirements of ISO 9001:2000. It is internationally accepted practice for Laboratories to obtain accreditation to ISO/IEC 17025, and not ISO 9001 certification, since obtaining ISO 9001 certification for Laboratories does not demonstrate technical competence to undertake scientific tests or measurements and issue reports that can be relied on. On the other hand, care has been taken to incorporate all the elements of ISO 9001:2000 relevant to the scope of testing and calibration services covered by the Laboratory's management system (Panadda, 2007).

The specific standard for medical Laboratories (ISO 15189) was published in 2003 to address the unique nature of medical Laboratories (Aoyagi, 2004) compared with other types of Laboratories, especially the pre-analytic and post-analytic parts of the quality system, as these two parts play a vital role in generating the results of tests. Moreover, the concept of "patient care" has also been emphasized in this new standard. The first edition of ISO 15189:2003 specifies requirements for quality and competence particular to medical Laboratories. It is based upon ISO/IEC 17025:1999 and ISO 9001. It is intended for use throughout the currently recognized disciplines of medical Laboratory services; those working in other services and disciplines could also find it usual and appropriate. The second edition of this international standard is aimed at more closely aligning with a second edition of ISO/IEC 17025:2005 and ISO 9001:2001 (Panadda, 2007).

The ISO 15189:2003 is widely used by all accreditation bodies throughout the world and accepted as the international standard for medical Laboratories. However, any standard needs to be reviewed periodically. The new version of ISO 15189:2007 has been published and is implemented by all accreditation bodies in two years period (Panadda, 2007). In the SEA Region by 2007, India, Indonesia and Thailand had established accreditation systems (**Table 3**) while other countries were still in the planning phase. In Sri Lanka, a 10-year plan for granting accreditation was developed by the Ministry of Science and Technology using ISO 15189 (Panadda, 2007).

Table 3. Status of accreditation in some countries in SE Asia*

Country	Quality system or accreditation agency	Membership, basis of accreditation and constraints	Standard followed
India	The National Accreditation Board for Testing and Calibration Laboratories (NABL) established by the Government of India in 1994	**Membership**: Asia Pacific Laboratory Accreditation Cooperation (APLAC) and International Laboratory Accreditation Cooperation (ILAC). **Basis**: Voluntary **Constraints**: -Absence of a national Laboratory policy and national standards for accreditation, except for blood banks. -Obtaining accreditation is still voluntary and there is poor participation of public sector	Initially, ISO Guide 58 and ISO 17025, later adopted ISO standards for clinical Laboratories.
Thailand	The Bureau of Laboratory Quality and Standards (BLQS) **	**Membership**: APLAC and ILAC. **Basis**: The accreditation body follows the international standard ISO 17011.	ISO/International Electrotechnical Commission (IEC) 17025 and ISO 15189
Indonesia	The National Accreditation Committee (KAN) and The Center for Health Laboratory		

Sources: *Panadda Silva, 2007. Guidelines on Establishment of Accreditation of Health Laboratories. WHO Regional Office for South-East Asia, New Delhi, 49 p. **WHO publication Quality standards in health Laboratories, implementation in Thailand: a novel approach, World Health Organization Regional Office for South-East Asia, 2005 (SEA-HLM-386)

Benefits of Laboratory accreditation

Laboratory accreditation is accompanied by a number of advantages, some of which are as follows (Panadda, 2007):

- Facilitates the implementation and maintenance of an effective quality system;
- Gives confidence to the Laboratory and to users for the results generated;
- Provides national and international acceptance of results and recognition of technical competence;
- Helps to defend Laboratories in case of legal disputes pertaining to Laboratory results;
- Reduces the operating costs of the Laboratories by getting correct results the first time and always;
- Helps private sector Laboratories to attract more business;
- Helps to meet purchase or regulatory specifications;
- Increases competitiveness and market share.

Chapter 10

Ethical issues in research management

So far, we have not dealt widely on all the various scientific disciplines that have to do with research and Laboratory operations within the framework of this book. Most of our viewpoints might seem to focus more on research in biosciences. This is mindful of the fact that intensive Laboratory work is undertaken in the biosciences, most of which are focused on issues related to human and animal health, food security and environmental concerns. In fact, the bioscience disciplines are mostly experimental sciences which have the following defining characteristics (Heather and Edward, 2005):

- Bioscience is considered as a family of methods and disciplines grouped around the investigation of life ;
- It deals with processes and the interrelationships of living organisms ;
- In biosciences, most or all knowledge is related to observation or experiment ;
- Bioscience disciplines exist in an environment of current hypotheses rather than certainty ;
- they include disciplines in which rapid change is happening ;
- they are essentially practical and experimental subjects ;

Ethical issues in biosciences which entail responsible, conscious and voluntary conduct (Redi, 2008), are a growing concern for companies in the wake of a series of corporate governance scandals and the accompanying decline in societal and investor trust in firms (Mackie *et al.,* 2006). In fact, scientific ethics facilitates free and accurate information transfer among stakeholders in research. In the domain of biosciences, major ethical elements include increased understanding of biological systems, responsible use of technology, and curtailment of ethnocentric debates more in tune with new scientific insights.

Coined by Irina Pollard in 1994, bioscience ethics is today an internationally recognized discipline, interfacing science and bioethics within professional perspectives such as medical, legal, bioengineering and economics (Irina Pollard, 2009). Bioscience ethics or bioethics aims at incorporating current scientific insights of practical significance within a cultural context while serving as the interface which ensures that such scientific information is not omitted or corrupted in this process (Irina Pollard, 2009, p. 22). The consideration of ethics in research should enhance mature decision-making in harmony with changing technology. Effective policies that guarantee all stakeholders' access to the benefits of the new bioscience technologies must be developed; such should be policies that are based on ethical principles which respect the plurality of values of the stakeholders, and which allow the development of homogenous regulations (Redi, 2008).

In fact, moral distress occurs when one knows the ethically appropriate action to take but is constrained from taking that action (Hamric *et al.,* 2011). Therefore, companies and Laboratories in the pharmaceutical, biotech, and bio-agricultural industries must not only comply with a wide array of government regulations to balance the profit motive with social responsibility, but also must deal with the complex array of ethical issues raised by doing business in the biosciences. Some of the ethical issues raised are as follows (Mackie *et al.,* 2006):

- the production and sale of genetically modified foods;
- animal testing for pharmaceuticals;
- drug pricing at home and in developing countries;
- the creation of transgenic animals for drug production;
- the potential misuse of personal genetic information;

- appropriate commercialization and profit making from genetic and biological samples;
- gene therapy experiments and embryonic stem cell research to produce new therapies.

One critical aspect of research ethics is the use of animal models. Animals play an important part in biomedical research, and much discussion concerns the ethical considerations of their use in research. Most published biomedical research involving animal models are regulated at local or national levels to ensure that appropriate ethical standards are met (Rands, 2011). The bioscience community accepts that animals should be used for research only within an ethical framework (Festing and Wilkinson, 2007).

The use of genomic databases also raises concern on informational risks to data subjects. Many human genome research databases contain large quantities of sensitive medical and other personal information. Some data elements in genome research databases might be similar or identical to elements in non-research databases, and might be used to link research databases with commercial, forensic, or other databases (Ossorio, 2011).

Biobanks constitute a significant research tool that has gained support from both the scientific community and regional, national and international research funding agencies (Caulfield *et al.*, 2014). However, their use (of biobanks) may create concerns about whether appropriate ethics oversight mechanisms are in place to deal adequately with a variety of commercialization-related concerns. Therefore, Biobanks management and commercialization processes must be done with consideration of policy challenges for scientists, research participants, and funders relative to ethical and legal norms (Caulfield *et al.*, 2014).

Some very important ethical issues in Laboratory research concern DNA technologies and research involving material transfer between Laboratories and research institutions. Serious worldwide controversies arose regarding potential risks and benefits as well as possible hazards associated with the recombinant DNA technologies (Gura, 1999; Cardoso *et al.*, 2005) after the first DNA experiments carried out in 1972 (Cohen *et al.,* 1972). The Asilomar Conference in California [1975] responded to the concerns by establishing clear rules for the control of these risks and setting up the ethical criteria and operational procedures necessary to ensure that any experiments in the field of genetic engineering is ethical and safe (Berg *et al.,* 1975; Cardoso *et al.,* 2005). For instance, with the publication of the human genome as a landmark in the transition of biology from essentially laboratory-based analysis to computer-driven analysis, or bioinformatics, Biotech Laboratories and pharmaceutical companies must operate on strict ethical basis in providing investment incentives for continued research and development (Irina Pollard, 2009).

But standing on ethical grounds in business is not that easy and requires courage and moral! Relating to controversy in the views of scientists and politicians whether there was a proof or not of a link between CJD (Creutzfeldt Jakob disease) and BSE (bovine spongiform encephalopathy), Irina Pollard (2009) points out that it is insulting and also potentially disastrous when the public is 'reassured' rather than informed about uncertainties, stating that "People have a right to the facts, and should develop a critical understanding of the world in which they live". According to this author, scientific knowledge grows in a climate of balanced acceptance of uncertainties, the need to preserve what has been proven and tested, and the courage, when the environment changes, to adjust to new realities (Irina Pollard, 2009, p. 22).

Some strategies to implement ethics

There is a variety of approaches by which research institutes or business companies of all sizes and in different market niches use to encourage ethical behavior. First of all, executives are promoting ethics as part of a firm's core values (Novas, 2006). As ethical concerns increase, some companies tend to retool their organizational structures with consideration of ethical values; ethics shape their hiring and staff performance evaluations, ethics training is given to employees in some firms, and visual and oral reminders are used in the workplace to reinforce the organizations' commitment to ethical values (Novas, 2006). One other approach to ensuring ethical values by companies is the hiring practices focused on ethics (Mackie *et al.*, 2006) through which some of the small- and medium-sized companies put weight on candidates' values in addition to their past performance and technical expertise, when making hiring decisions, with the aim of assessing how the potential employee's values align with the ethical values of the company (Mackie *et al.*, 2006). They actually employ individuals, including key business leaders who include ethics as part of their job responsibilities (Mackie *et al.*, 2006).

Several of the larger companies develop separate departments like an internal Ethics Office with a Bioethics Director whose sole focus is to address the company's environmental, human, and animal-related ethical issues (Mackie *et al.*, 2006; Novas, 2006), while some smaller firms incorporate ethics into the responsibilities of senior managers (Novas, 2006). Where internal expertise is lacking, external consultants are brought in or independent ethics advisory boards are created to provide guidance, ethics education and advice at various stages in their development (Mackie *et al.*, 2006; Novas, 2006).

A further approach is External Ethics Engagement through which companies develop ethics mechanisms with their partners and suppliers in order to ensure that their partners also follow high ethical standards. For instance, some companies devise a mechanism with countries that are involved with the collection of biological samples. Instead of secretly taking genetic material from these countries (biopiracy), they form partnerships with them involving collection and processing of samples, funding and training, along with a royalty percentage on any discoveries that originate from the samples (Mackie *et al.*, 2006).

Material transfer and ethics

Almost all Laboratories undertake research involving the transfer of materials between individuals or institutions. The materials may include reagents, cell lines, antibodies, genome sequence databases, vectors, live animals, transgenic animals, seeds, live plants or plant extracts, chemical compounds like drugs or pharmaceuticals and accompanying information and data. From an ethical standpoint, it is imperative to respect existing norms/standards, laws and regulations including export and import permits relative to such transactions. For instance, ensuring that a phytosanitary certificate accompanies samples in the course of their transfer between researchers is a basic ethical principle. Moreover, the establishment of material transfer agreements (MTA) between parties involved in such exchanges of materials is good practice.

An MTA which involves the transfer of possession and not legal title is a legal agreement between the provider (owner) and the recipient of research materials, for testing, evaluation, experimentation, or research. The MTA serves to protect the rights of the parties while allowing research, analysis and evaluation to proceed unhindered. Well conceived and established MTAs also protect the rights of researchers to publish research results and participate in ownership of any eventual Intellectual Property (IP). They also protect the providing and receiving organizations from misunderstandings and potential liability. Proper baseline negotiation involving all stakeholders is necessary at all stages of MTA preparation, establishment and implementation.

It is not possible to treat all ethical issues related to research within the framework of this write-up. However, more information on ethical issues and useful links related to bioscience and bioethics can be accessed at the Bioscience-Bioethics Friendship Co-operative (BBFC) web portal at http: //www.bioscience-bioethics.org/.

Conclusion

This book is our modest contribution to efficient Laboratory management. It would have been pretentious to say it contains all the wealth of knowledge needed to fully run a modern Laboratory or research facility efficiently. However, our wish is that it bridges the gaps in the development and implementation of policies and practices related to Laboratory management, especially regarding Biosafety and Biosecurity. Through this book, we hope that Laboratory operators will begin by changing personal attitude at the work place. Finally, we recommend that for the smooth running of a Laboratory, the Managers should make handy or implement the following:

- A complete and regularly updated list of equipment and supplies of the Laboratory;
- Standard Operation Procedures or protocols for all equipment in the Laboratory;
- Produce operation manuals for all experiments undertaken in the Laboratory;
- Develop a floor and equipment setting plan of the Laboratory for safe access and operations;
- Establish and effectively monitor a procurement program for Laboratory materials and equipment;
- Educate other Laboratory users on Laboratory management principles;
- Purchase or produce a customized LIMS for the Laboratory;

- Develop a process map that outlines all the steps of the testing processes underway in the Laboratory to facilitate risk assessment and enhance safety;
- Above all, Laboratory staff should collaborate efficiently and convincingly with the hierarchy and decision making bodies!!!

References

1. Abad Morejon de Giron, Solanes Foz, Domingo Alvarez, 2012. Reflections on Biosafety: do we really know what Biosafety, biocontainment, and Biosecurity mean? *Contributions to Science*, 6 (1): 99–103.
2. Alejandro M. C., 2011. Total Quality Management: Quality Culture, Leadership and Motivation. Master Thesis: Total Quality Management, Politecnico di Milano, 94pp.
3. American Biological Safety Association (ABSA), 2003. Available at www.absa.org.
4. Anonymous, 2012. Laboratory information management. http:/en.wikipedia.org/wiki/Laboratory-information-management (accessed on 5th November 2012).
5. Aoyagi T., 2004. ISO 15189 medical Laboratory accreditation. *Rinsho Byori*, 52 (10): 860- 65.
6. Arnold D. K., Curtis P. McLaughlin, and Kit Simpson, 1992. Applying Total Quality Management Concepts to Public Health Organizations. *Publ. Health Reports* (May-June), 107 (3) : 257-264.
7. Bakanidze L, Imnadze P., and Perkins D., 2010. Biosafety and Biosecurity as essential pillars of international health security and cross-cutting elements of biological nonproliferation. *BMC Public Health*, 10 (Suppl 1): S12. 8 p. doi:10.1186/1471-2458-10-S1-S12
8. Bondad-Reantaso, M. G., Arthur, J. R. and Subasinghe, R. P., eds. 2012. Improving Biosecurity through prudent and responsible use of veterinary medicines in aquatic food production. FAO Fisheries and Aquaculture Technical Paper. No. 547. Rome, FAO. 207 pp.

9. Brian B. Mansir and Nicholas R. Schacht, 1989. Total Quality Management: A guide to implementation. LMI Report PL912R1. Logistics Management Institute, Goldsboro Road Bethesda, Maryland, 137pp.
10. Cardoso T. A. O., Albuquerque Navarro M. B. M., Soares B. E. C., Lima e Silva F. H., Rocha S. S., and Oda L. M., 2005. Memories of Biosafety in Brazil: Lessons to be learned. *Applied Biosafety*, 10 (3): 160-168.
11. Caulfield T., Burningham S., Joly Y., Master Z., Shabani M., Borry P., Becker A., Burgess M., Calder K., Critchley C., Edwards K., Fullerton S. M., Gotweis H., Hyde-Lay R., Illes J., Isasi R., Kato K., Kaye J., Knoppers B., Lynch J., McGuire A., Meslin E., Nicol D., O'Doherty K., Ogbogu U., Otlowski M., Pullman D., Ries N., Scot C., Sears M., Wallace H. and Zawati M. H., 2014. A review of the key issues associated with the commercialization of biobanks. *Journal of Law and the Biosciences*, 94-110. DOI:10.1093/jlb/lst004
12. CDC and NIH, 1984. Centers for Disease Control and Prevention (CDC) and National Institutes of Health (NIH). Biosafety in microbiological and biomedical Laboratories, p. 100. Washington, DC.
13. CDC and NIH, 1999. Centers for Disease Control and Prevention (CDC) and National Institutes of Health (NIH). Biosafety in microbiological and biomedical Laboratories, p. 250. Washington, DC: Author.
14. Clifton R. Lacy, M.D., Albert G. Kroll, and James E. McGreevey, 2003. Personal Protective Equipment. Public Employees Occupational Safety and Health Program. New Jersey Department of Health and Senior Services and New Jersey Department of Labor. www.state.nj.us/health/eoh/peoshweb. NJDHSS, 4pp.
15. Dargahi H. and Rezaiian M., 2007. Correlation between Knowledge, Attitude and Performance of the Employees with Quality Assurance System Implementation by the Employers. *Iranian J Publ Health,* 36 (3): 45-51.

16. Devendra T. Mourya, Pragya D. Yadav, Triparna Dutta Majumdar, Devendra S. Chauhan* and Vishwa Mohan Katoch, 2014. Establishment of Biosafety Level-3 (*BSL-III*) Laboratory: Important criteria to consider while designing, constructing, commissioning and operating the facility in Indian setting. Special report, *Indian J. Med. Res.* (August 2014), 140: 171-183.
17. Di Benedetto A., Dolcetti G., Grimaz S., Russo P. and Salzano E., New Trend In Classification Of Hazard Of Explosive Atmosphere In Laboratory Workplaces. 6p.
18. Drosten C., Günther S., Preiser W. *et al.,* 2003. Identification of a novel coronavirus in patients with severe acute respiratory syndrome. *New England Journal of Medicine*, 348: 1967-1976.
19. Ezzelle J., Rodriguez-Chavez I. R., Darden J. M., Stirewalt M., Kunwar N., Hitchcock R., Walter T., and D'Souza M. P., 2008. Guidelines on Good Clinical Laboratory Practice: Bridging Operations between Research and Clinical Research Laboratories. *J. Pharm. Biomed. Anal.* (January 7); 46 (1): 18-29.
20. FAO, 2013. Quality assurance for microbiology in feed analysis laboratories. FAO animal production and health manual No. 16. Rome. p. 23.
21. Festing S. and Wilkinson R., 2007. The ethics of animal research. *EMBO reports*, 8 (6): 526-530.
22. Gaudioso J., and Zemlo T., 2007. Survey of Bioscience Research Practices in Asia: Implications for Biosafety and Biosecurity. *Applied Biosafety,* 12 (4): 260-267.
23. Gaudioso J., Rivera S. B., Caskey S., and Salerno R. M., 2006. Laboratory Biosecurity: A Survey of the U.S. Bioscience Community. *Applied Biosafety*, 11 (3): 138-143.
24. Geir T., Jevnaker M., Guri A. G., and Sandberg S., 2011. Quality assurance of laboratory work and clinical use of laboratory tests in general practice in Norway: A survey. *Scandinavian Journal of Primary Health Care*, 29: 171–175. DOI: 10.3109/ 02813432.2011.585837

25. Gray J. J., T. Wreghitt G., McKee T. A., McIntyre P., Roth C. E., Smith D. J., Sutehall G., Higgins G., Geraghty R., Whetstone R., Desselberger U., 1995. Internal quality assurance in a clinical virology laboratory. II. Internal quality control. *J. Clin. Pathol.* 48: 198-202.
26. Guangshu C., 2009. Total Quality Management in Supply Chain. *International Business Research*, 2 (2): 82-85.
27. Hamric A. B., Faan R. N., Borchers T. C., Epstein E. G., 2011. Moral distress and ethical climate in nurses and physicians in intensive care unit (ICU) settings. Poster: Presidential Inauguration Research Poster Competition, University of Virginia, April 14, 2011.
28. Heather J. S. and Edward J. W., 2005. Linking Teaching and Research in the Biosciences. BEE-j Volume 5: May 2005. Online at http://www.bioscience.heacademy.ac.uk/journal/vol5/beej-5-4.pdf
29. Herman P., Verlinden Y., Breyer D., Van Cleemput E., Brochier B., Sneyers M., Snacken R., Hermans P., Kerkhofs P., Liesnard C., Rombaut B., Van Ranst M., van der Groen G., Goubau P., and Moens W., 2004. Biosafety Risk Assessment of the Severe Acute Respiratory Syndrome (SARS) Coronavirus and Containment Measures for the Diagnostic and Research Laboratories. *Applied Biosafety*, 9 (3): 128-142.
30. Houston A. and Dockstader S. L., 1988. A Total Quality Management Process Improvement Model. Navy Personnel Research and Development Center. San Diego, CA 0'2152-68W NPRDC TR 89-3, December 1988, 73pp.
31. Hruz P., 2008. Essential Laboratory Management Skills (Lecture), 11 p.
32. HSE, 1992. A short guide to the Personal Protective Equipment at Work, Regulations 1992. A web-friendly version of leaflet INDG174 (rev1), revised 08/05. 5pp.
33. HSE, 2012. Personal protective equipment (PPE): Equipment and method sheet, em6 asbestos essentials-Nonlicensed

tasks. Published by the Health and Safety Executive 04/12. 3pp.

34. Ingrid Schmid, 2012. How to develop a Standard Operating Procedure for sorting unfixed cells. Methods 57(3): 392–397. doi:10.1016/j.ymeth.2012.02.002.
35. Ippolito G., Nisii C., Di Caro A., Brown D., Gopal R., Hewson R., Lloyd G., Gunther S., Eickmann M., Mirazimi A., Koivula T., Georges Courbot M.-C., Raoul H., and Capobianchi M.R., 2009. European Perspective of 2-Person Rule for Biosafety Level 4 Laboratories. Emerging Infectious Diseases (Letters) www.cdc.gov/eid, 15 (11): 1858-1860, November 2009.
36. Irina Pollard, 2009. Bioscience ethics. Macquarie University, New South Wales, Australia. Published in the United States of America by Cambridge University Press, New York.
37. Isin Akyar, 2011. GLP: Good Laboratory Practice, Modern Approaches To Quality Control, Dr. Ahmed Badr Eldin (Ed.), ISBN: 978-953-307-971-4, InTech, Available from: http://www.intechopen.com/books/modernapproaches-to-quality-control/glp-good-Laboratory-practice (16/11/2014)
38. ISO 15189, International Organization for Standardization. ISO 15189: Medical laboratories, particular requirements for quality and competence. 2003. http://www.iso.org/iso/en/CatalogueDetailPage.CatalogueDetail?CSNUMBER=26301
39. Jeffrey A. M. and Stefanie E. N., 2009. Research trends in the Academy of Management publications. *Journal of Management and Marketing Research. Research Trends*, 1-31.
40. Kanagasabapathy A. S. and Pragna R., 2005. Laboratory Accreditation - Procedural Guidelines. *Indian Journal of Clinical Biochemistry*, 20 (2): 186-188.
41. Ksiazek, T. G., Erdman, D., Goldsmith, C. S., *et al.*, 2003. A novel coronavirus associated with SARS. *New England Journal of Medicine*, 348, 1953-1966.
42. Kubono K., 2004. Quality Management System in the medical Laboratory. ISO 15189 and Laboratory accreditation. *Rinsho Byori*, 52 (3): 274-78.

43. Le Duc J. W., Anderson K., Bloom M. E., Carrion R. Jr., Feldmann H., Fitch J. P., *et al.,* 2014. Potential impact of a 2-person security rule on Biosafety Level 4 Laboratory workers. Emerg Infect Dis [cited 2014 Oct 20]. Available from http://www.cdc.gov/EID/content/15/7/e1.htm DOI: 10.3201/eid1507.081523.
44. Leunda A, Roels S, Van Vaerenbergh B., 2011. Dismantlement of Laboratories performing rapid detection of Transmissible Spongiform Encephalopathy. Report: ISP/41/CU/11-1034. http://www.Biosafety.be/CU/PDF/Dismantlement_Lab_TSE.pdf
45. Leunda A., Van Vaerenbergh B., Baldo A., Roels S., and Herman P., 2013. Laboratory activities involving transmissible spongiform encephalopathy causing agents. Risk assessment and Biosafety recommendations in Belgium. Technical Report. Prion (Landes Bioscience), 7 (5): 420-433.
46. Mackie J. E., Taylor A. D., Finegold D. L., Daar A. S., Singer P. A., 2006. Lessons on ethical decision making from the bioscience industry. *PLoS Med,* 3(5): e129. DOI: 10.1371/journal.pmed.0030129
47. Marya L. Wilson, 2006. Total Quality Management (TQM) at the University Centers. A Research Paper Submitted in Partial Fulfillment of the Requirements for the Master of Science Degree in Management Technology. The Graduate School, University of Wisconsin-Stout, Menomonie, WI 5475, 53pp.
48. Mikulich V. J., Schriger D. L., 2002. Abridged version of the updated US Public Health Service guidelines for the management of occupational exposures to hepatitis B virus, hepatitis C virus, and human immunodeficiency virus and recommendations for postexposure prophylaxis. *Ann. Emerg. Med.*; 39:321–328.
49. Mothabeng D, Maruta T, Lebina M, Lewis K, Wanyoike J, Mengstu Y., 2012. Strengthening Laboratory Management Towards Accreditation: The Lesotho experience. *Afr J Lab*

Med., 1(1), Art. #9, 7 pages. http://dx.doi. org/10.4102/ajlm. v1i1.9

50. Nema, A., Pathak, A., Bajaj, P., Singh, H., and Kumar, S., 2011. A case study: biomedical waste management practices at city hospital in Himachal Pradesh. *Waste Management and Research*, 29(6), 669-673.
51. Nichols J. H. 2011. Laboratory quality control based on risk management. *Annals of Saudi Medicine*, 31(3): 223-228. doi: 10.4103/0256-4947.81526
52. Noori H., 2004. Collaborative Continuous Improvement Programs in Supply Chain. *Problems and Perspectives in Management*, 2: 228-245.
53. Novas C., 2006. What is the bioscience industry doing to address the ethical issues it faces? *PLoS Med*, 3(5): e142. doi: 10.1371/journal.pmed.0030142
54. OECD, 1998. OECD series on Principles of Good Laboratory Practice and Compliance Monitoring. Available on the Internet at: http://www.oecd.org/officialdocuments/displaydocumentpdf/?cote=env/mc/chem(98)17anddoclanguage=en
55. Orgam D., 1995. Quality systems for the clinical Laboratories. Canadian Society of Laboratory Technologies Working Group. *Can J Med Technal*, 57 (2): suppl 1-14.
56. OSHA, 2000. Assessing the Need for Personal Protective Equipment: A Guide for Small Business Employers. Small Business Safety Management Series, U.S. Department of Labor, Occupational Safety and Health Administration (OSHA 3151). 57pp.
57. Ossorio P. N., 2011. Bodies of Data: Genomic Data and Bioscience Data Sharing. *Soc. Res.* (New York), 78(3): 907-932.
58. Panadda S., 2007. Guidelines on Establishment of Accreditation of Health Laboratories. WHO Regional Office for South-East Asia, New Delhi, 49 p.
59. Peiris J. S. M., Lai S. T., Poon L. L. M. *et al.*, 2003. Coronavirus as a possible cause of severe acute respiratory syndrome. *Lancet*, 361: 1319-1325.

60. Perigo D. and Rabelo R., 2005. ISO- 2: Diving deeper into ISO 9001-2000 and the role of quality management systems in clinical Laboratories. Online at: www.westgard.com.
61. Phillips J. A., 1998. Managing America's Solid Waste. J.A. Phillips and Associates Boulder, Colorado. NREL/SR-570-25035, 172 p.
62. Phu Van Ho, 2011. Total Quality Management Approach to the Information Systems Development processes : An empirical study. Ph.D. Thesis, Faculty of Virginia Polytechnic Institute and State University, Alexandria, Virginia, 297pp.
63. Rabinson S., 1999. Measuring services quality: Current thinking future requirement. *Marketing and Planning*, 17(1): 210-24.
64. Rands S. A, 2011. Inclusion of policies on ethical standards in animal experiments in biomedical science journals. *Journal of the American Association for Laboratory Animal Science*, 50 (6): 901-903.
65. Redi C. A., 2008. The problems of biosciences in contemporary society. *Contributions to Science*, 4 (1): 97-104. doi: 10.2436/20.7010.01.4
66. Royce and John R., 2010. "Industry Insights: Examining the Risks, Benefits and Trade-offs of Today's LIMS". Scientific Computing. Retrieved 13 February 2012.
67. Sabelnikov A., Zhukov V., and Kempf R., 2006. Some bioterrorism issues of quantitative Biosafety. *Applied Biosafety*, 11(2): 67-73.
68. Seiler J. P., 2005. Good Laboratory Practice. The why and the how. ISBN 3-540-25348-3, Springer-Verlag Berlin Heidelberg, Printed in the European Union.
69. Simpson K. N., Kaluzny A. D., and McLaughlin C. P., 1991. Total quality and the management of laboratories. *Clin. Lab. Manage. Rev.* (Nov-Dec), 5 (6) : 448-9, 452-3, 456-8, passim.
70. Stanford LSDG, 2013. This module has been adapted from the Stanford Laboratory Standard and Design Guide (Standford LSDG); General Requirements for Stanford Laboratories Version 2.0/ 11-06, 13 p.

71. Stavskiy E. A., Barbara J., Robert J. Hawley, Jonathan T. Crane, Nikolay B. Cherny, Irina V. Renau, and Sergey V. N., 2003. Comparative analysis of Biosafety guidelines of the USA, WHO, and Russia (Organizational and controlling, medical and sanitary-antiepidemiological aspects). *Applied Biosafety*, 8 (3): 118-127.
72. Summermatter K., 2009. Biosafety-Europe: Coordination, harmonization and exchange of Biosafety and Biosecurity practices within a Pan-European Network. *Applied Biosafety*, 14 (2): 105.
73. Tabasi R. and Marthandan G., 2013. Clinical Waste Management: A Review on Important Factors in Clinical Waste Generation Rate. *International Journal of Science and Technology*, 3 (3) : 194-200.
74. Tazawa H., 2004. The College of American Pathologists (CAP) quality management system in clinical Laboratory and its issue. *Rinsho Byori*, 52 (3): 266-69.
75. Tun T., Sadler K. E., and Tam J. P., 2007. Biological agents and toxins Act: Development and enforcement of Biosafety and Biosecurity in Singapore. *Applied Biosafety*, 12 (1): 39-43.
76. UNEP, 1996. United Nations Environment Programme (UNEP). International technical guidelines for safety in biotechnology.
77. Wang SA, Panlilio AL, Doi PA, White AD, Stek M Jr. Saah A., 2000. Experience of healthcare workers taking postexposure prophylaxis after occupational HIV exposures: findings of the HIV Postexposure Prophylaxis Registry. *Infect. Control Hosp. Epidemiol.*, 21:780–785.
78. Westgard J. O., Barry P. L., and Tomar R. H., 1991. Implementing total quality management. *Clin Lab Manage Rev.* 1991 Sep-Oct;5 (5) :353-5, 358-9, 362-6 passim.
79. WHO, 2003. World Health Organization Laboratory Biosafety manual. Available at http://who.int/entity/csr/resources/publications/Biosafety/en.
80. Xin Pan, 2012. Isolator System for Laboratory Infectious Animals, Insight and Control of Infectious Disease in Global

Scenario, Dr. Roy Priti (Ed.), ISBN: 978-953-51-0319-6, InTech, Available from: http://www.intechopen.com/books/insight-and-control-of-infectious-disease-in-global-scenario/isolator-systemfor-Laboratory-infectious-animals.

81. Yeniseis O. P., *2008*. Good Laboratory Practices and the ISO 9001:2000 standards. *Biotecnología Aplicada, 25: 258-261.*

www.ingramcontent.com/pod-product-compliance
Lightning Source LLC
Chambersburg PA
CBHW030840180526
45163CB00004B/1394
9781503564374